"ASPECT" GEOGRAPHIES

A GEOGRAPHY OF AGRICULTURE

PAUL A. R. NEWBURY
B.Sc. (Econ.), B.Phil., Dip. Ed., F.R.G.S.

Lecturer in Geography, Dept. of General Education and Science, Milton Keynes College,
formerly Head of the Geography Department,
Eastwood Hall Park Technical Grammar School, Nottinghamshire

MACDONALD AND EVANS

MACDONALD & EVANS LTD
Estover, Plymouth PL6 7PZ

First published 1980
Reprinted 1982

©
MACDONALD & EVANS LTD
1980

ISBN 0 7121 0733 9

Other "Aspect" Geographies
BIOGEOGRAPHY
A GEOGRAPHY OF MANUFACTURING
GEOGRAPHY OF SETTLEMENTS
GEOGRAPHY OF TRANSPORT
TROPICAL GEOGRAPHY

*Printed in Great Britain by
Butler & Tanner Ltd.
Frome and London*

Introduction to the Series

THE study of modern geography grew out of the medieval cosmo-graphy, a random collection of knowledge which included astronomy, astrology, geometry, political history, earthlore, etc. As a result of the scientific discoveries and developments of the seventeenth and eight-eenth centuries many of the component parts of the old cosmography hived off and grew into distinctive disciplines in their own right as, for example, physiography, geology, geodesy and anthropology. The residual matter which was left behind formed the geography of the eighteenth and nineteenth centuries, a study which, apart from its mathematical side, was encyclopaedic in character and which was purely factual and descriptive.

Darwinian ideas stimulated a more scientific approach to learning, and geography, along with other subjects, was influenced by the new modes of thought. These had an increasing impact on geography, which during the present century has increasingly sought for causes and effects and has become more analytical. In its modern development geography has had to turn to many of its former offshoots—by now robust disci-plines in themselves—and borrow from them: geography does not attempt to usurp their functions, but it does use the material to illuminate itself. Largely for this reason geography is a wide-ranging discipline with mathematical, physical, human and historical aspects: this width is at once a source of strength and weakness, but it does make geography a fascinating study and it helps to justify Sir Halford Mackinder's con-tention that geography is at once an art, a science and a philosophy.

Naturally the modern geographer, with increasing knowledge at his disposal and a more mature outlook, has had to specialise, and these days the academic geographer tends to be, for example, a geomorpho-logist or climatologist or economic geographer or urban geographer. This is an inevitable development since no one person could possibly master the vast wealth of material or ideas encompassed in modern geography.

This modern specialisation has tended to emphasise the importance of systematic geography at the expense of regional geography, although it should be recognised that each approach to geography is incomplete without the other. The general trend, both in the universities and in the school examinations, is towards systematic studies.

This series has been designed to meet some of the needs of students pursuing systematic studies. The main aim has been to provide in-troductory texts which can be used by sixth-formers and first-year uni-versity students. The intention has been to produce readable books which will provide sound introductions to various aspects of geography, books which will introduce the students to new ideas and concepts as well as more detailed factual information. While one must employ precise scientific terms, the writers have eschewed jargon for jargon's sake; moreover, they have aimed at lucid exposition. While, these days, there is no shortage of specialised books on most branches of geographical knowledge, there is, we believe, room for texts of a more introductory nature.

The aim of the series is to include studies of many aspects of geography embracing the geography of agriculture, the geography of manufacturing industry, biogeography, land use and reclamation, food and population, the geography of settlement and historical geography. Other new titles will be added from time to time as seems desirable.

H. ROBINSON
Geographical Editor

Preface

As is more fully elaborated elsewhere, it has been the aim of the author to present a geography of agriculture, at once different from and complementary to the majority of others. It needs to be "different" because, with the current preoccupation with educing abstract and quantified information, a trend has developed for ousting the concrete sample studies which could afford an essential insight into the quality of phenomena under consideration. It is, however, intended that the book should be "complementary", attempting not to reverse the trend but rather to adjust the imbalance, enabling students to acquire a more over-all appreciation of their subject.

To this end, the author has searched out and consulted agricultural authorities and practising farmers all over the world. This has proved a difficult task, inevitably involving much effort, frustration and disappointment, and the result is the product of some compromise. The work has also demanded much time and effort on the part of an obliging group of collaborators. Every effort has been made to acknowledge such contributors in situ, but the list below also includes those whose support may have been less specific, but no less essential to the production of the book. To these and any others, inadvertently omitted, sincerest thanks are due.

Invaluable help has been received from the following: Kathleen Taylor, Botswana; David Pollard, Goondiwindi, Queensland; the late Gwynedd Pritchard of Maentwrog, North Wales; Messrs. George Insko, B. Tilson, Oscar Penn, Clarence Lebus and William Lovell Hollar in Kentucky; John and Veronica Eastabrooks of Cincinnati; Dr. Karl B. Raitz of the University of Kentucky Geography Department; B. W. Gunn, Corporate Secretary of C.S.P. Foods, Saskatoon; Messrs. Warren Loyns and Stan Hicks of the Lac Vert and Ninga Prairie Farms respectively; the Rubber Research Institute of Malaysia; H. M. Collier of Dunlop Estates, Berhad; G. M. McManaman of Kuala Jelei Estate; Jon and Gunnar Øyro along with Messrs. Nes, Hus and Mehl in Rosendal, Norway; Klaas and Liesbeth Blokker along with Messrs. Vermeer, Op 't Hoog, Simons, van Alphen and van Kempen in Haaren, North Brabant, the Netherlands; Michael Nokes of Linslade Manor Farm; Miss Engee Caller and Mrs. Dvorah Shlossberg of Kibbutzim Kfar Blum and Gesher Haziv, Israel; Jim W. Parker of the L.S.A. Estate, Potton, Bedfordshire; F. H. Webster and the staff of the Plunkett Foundation for Co-operative Studies in Oxford; John Marangos of Bletchley, Milton Keynes, Dr. Brian Beeley of the South-East Region of the Open University in East Grinstead; and Professor Andrew

T. A. Learmonth, Head of the Geography Department of the Open University in Milton Keynes.

Finally, in undertaking this work I have been indebted to my students at Bletchley College of Further Education in Milton Keynes, through whose critical interest I have been able to develop my ideas and to my family and friends who have had to bear with my long preoccupation. In particular I owe so much to the unstinting support of my wife, Margo, "for encouragement, advice and assistance, and above all for the cheerful enthusiasm with which she endured the tribulations of driving a geographer in his never-satisfied search for geography" (quoted from J. Wrexford Watson's preface to his *North America—its Countries and Regions*, Longman, 1963).

March 1980

P. A. R. N.

Contents

Part One

INTRODUCTION

Chapter I

Theoretical Aspects of Geographical Location Relevant to Agriculture

THEORIES of geographical location owe much to the combined work of three Germans: von Thünen, Weber and Christaller, as well as to those who have subsequently developed and applied their ideas.

Of these, von Thünen was specifically concerned with agricultural land-use patterns. Christaller's main preoccupation was with settlement patterns, but in the context of agricultural regions his work is relevant to the probable hierarchy of market towns, whose related market catchments and service areas extend over the agricultural land. Finally, Weber's work, which was directed towards consideration of industrial location, can, as Chisholm points out in his *Rural Settlement and Land Use*,* be equally well applied to choices exercised by the farmer entrepreneur in newly colonised or reclaimed lands for selecting the most economical location for his farmhouse in relation to the probable inputs and outputs of his land, or for a land-use pattern consonant with its optimum profitability.

Von Thünen, his Work and its Significance

In 1826 Johann Heinrich von Thünen (1783–1850), in his classical book on agricultural land-use, *The Isolated State*, based his observations on accounts and records compiled in administering his own estates at Tellow, near Rostock, situated on the Baltic coast of Germany.

Although it is clear that agriculture has changed radically in the 150 years that have elapsed since von Thünen's day, and that in any case conditions over the earth's surface are variable, the basic soundness of his theoretical model has been testified to by geographical observers and locational theorists alike, who confirm its continued relevance not only in diverse times and situations, but also at scales ranging from the single farm unit to whole continents.

In this chapter von Thünen's original models, and some others adapted from or compatible with them, are considered. Full references for subsequent fuller enquiry are also provided.

Von Thünen's early model

For this early model (*see* Fig. 1(*a*)) von Thünen laid down the following postulates.

* *See* Bibliography for this and other books referred to in the text.

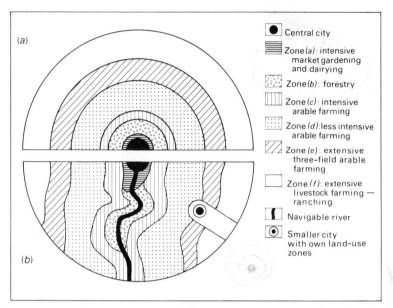

FIG. 1.—(a) *Von Thünen's early model of a single city in the centre of a homogeneous plain.* (b) *Von Thünen's later model modified by the addition of a smaller city and a single navigable river.*

1. An isolated city sits in the centre of a single homogeneous hinterland so that everywhere the fertility and therefore production costs, as well as ease of movement and therefore transport costs, are equal.
2. The city monopolises all the surplus produce of that hinterland, being its sole urban market.
3. Profit-motivated farmers are prepared to adapt their output in order to maximise their profits from this single market.
4. There is a single mode of transport the costs of which are borne by the farmer and are directly proportional to the distance to the city market.

Granted these, von Thünen asserted, a series of concentric zones, each with its own characteristic land-use, would be developed around the central city, as follows.

1. *Intensive market-gardening and dairying zone.* Milk and vegetables, because of demand, are commodities of high value, and so are produced close to markets to minimise costs incurred through their perishability.
2. *Forestry.* High demand for firewood, the essential fuel of the city, ensures its profitability; its bulk, which makes it costly to trans-

port, justifies the close proximity to the city market accorded to the forestry zone.

3. *Intensive arable.* Sixfold rotation, with three cereal crops and no fallow, means that land is 100 per cent tilled at any time. Cattle are stall-fed in winter.

4. *Less intensive arable.* Sevenfold rotation, with three cereal crops and one fallow, means that land is 86 per cent tilled at any time. Cattle are at pasture.

5. *Extensive three-field arable system.* Threefold rotation, with one cereal crop and one fallow, means that land is only 66 per cent tilled at any given time. Cattle are at pasture.

6. *Extensive livestock farming, or in this case, ranching.* Cattle are marketed "on hoof". Cheese is portable and sufficiently valuable to bear the high cost of transportation from a distance.

The extent of each zone would be determined by direct competition between products, based upon the formula:

$$V - (E + T) = P$$

where V is the value of the product, E the expenses of production, T is the cost of transport, and P the profit.

This process of land-use determination may best be illustrated by reference to the relative costs of wood and grain and how these affect their placement in relation to the city market, as shown in Table 1.

For each commodity, the outer production limit would occur where

TABLE 1

Relative costs of wood and grain at different distances from the city

Distance from city (km)	Wood $V - (E + T) = P$		Grain $V - (E + T) = P$	
1	$200 - (140 + 10\ ) =$	50	$80 - (50 + \ 3) =$	27
2	$200 - (140 + 20\ ) =$	40	$80 - (50 + \ 6) =$	24
3	$200 - (140 + 30\ ) =$	30	$80 - (50 + \ 9) =$	21
4	$200 - (140 + 40\ ) =$	20	$80 - (50 + 12) =$	18 (A)
5	$200 - (140 + 50\ ) =$	10	$80 - (50 + 15) =$	15
(B) 6	$200 - (140 + 60\ ) =$	0	$80 - (50 + 18) =$	12
7	$200 - (140 + 70\ ) =$	-10	$80 - (50 + 21) =$	9
8	$200 - (140 + 80\ ) =$	-20	$80 - (50 + 24) =$	6
9	$200 - (140 + 90\ ) =$	-30	$80 - (50 + 27) =$	3
10	$200 - (140 + 100) =$	-40	$80 - (50 + 30) =$	0 (C)

NOTES: Wood costs more than grain to transport because of its greater bulk. (A) is where inner limit of grain occurs because there is greater profit from wood. (B) is where the outer limit of wood occurs because profit has dwindled to nothing. (C) is where the outer limit of grain occurs because profit has dwindled to nothing.

increased cost of transportation absorbed profit; inner production limits would occur where greater profit was available from another commodity.

In Chapter V Ricardo's theory of economic rent will be considered. It will be demonstrated that von Thünen's model was based on the identical principle, although he arrived at it independently.

Students often complain that von Thünen's early model is unrealistic. In fact it was never meant to represent reality, but instead was intended to provide a sort of standard, so that when in subsequent model situations economic "variables" were introduced, these could be assessed as deviating from the first models.

Von Thünen's later model and further developments

In his later model (*see* Fig. 1(*b*)), von Thünen modified the conditions he had laid down as postulates of his earlier model, introducing a navigable river which, traversing all the zones on one side of the city, made possible lower transport costs than hitherto possible, particularly for those farmers whose land lay adjacent to it. This had the effect of re-aligning all the productive zones relating to the central city, elongating them to form narrow zones more or less parallel to the course of the river.

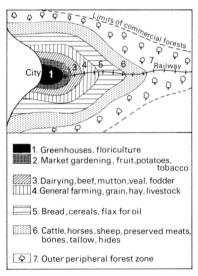

FIG. 2.—*Von Thünen's later model: zones of production about a theoretical isolated city in Europe.* (From "The agricultural regions of Europe", O. Jonasson, *Economic Geography*, Vol. I (1925), pp. 277–314.)

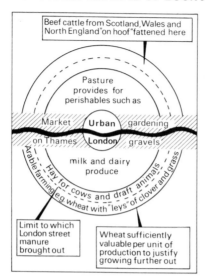

FIG. 3.—*Von Thünen's later model: zones of production around London, 1811.* (From the description of Rev. Henry Hunter's Model of London in *Rural Settlement and Land Use*, M. Chisholm.)

Similar situations are demonstrated by models developed by Jonasson (*see* Fig. 2), Hunter (*see* Fig. 3), and Nicolai and Jacques (*see* Figs. 4 and 5).

The zones shown in Fig. 2 were used for the following different types of agricultural production:

1. greenhouses, floriculture
2. market gardening, fruit, potatoes, tobacco   }—*horticulture*
3. dairying, beef, mutton, veal, fodder   }—*intensive farming*
4. general farming, grain, hay, livestock   *with dairying*
5. bread, cereals, flax for oil—*extensive agriculture*
6. cattle, horses, sheep, preserved meats, bones, tallow, hides   }—*extensive livestock*
7. outer peripheral forest zone—*forest culture*

In 1920, in the Congo, a 1,100 km railway was built from Port Franqui to Katanga, and by 1951 it had modified the distribution of population. Before 1920, this had been sparse and evenly spread, but by 1951 many one-time labourers had settled near the railway line (*see* Fig. 4) because of the enhanced market opportunities enjoyed near the line. This was also reflected in higher produce prices near the line (*see* Fig. 5).

Von Thünen went on to introduce an additional subsidiary city (*see*

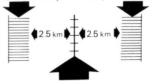

77% population on 95% land area, that is, more than one hour's walk from railway for man heavily laden with produce to sell

2.5 km 2.5 km

23% population in 5 km zone athwart railway line (5% total land area)

FIG. 4.—*Von Thünen's later model: effects of a railway on population distribution.* (From "La transformation des paysages congolais par le chemin de fer", H. Nicolai and J. Jacques, *Mémoires, Institut Royal Colonial Belge*, 1954.)

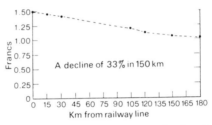

A decline of 33% in 150 km

Km from railway line

FIG. 5.—*Von Thünen's later model: the price of manioc related to distance from railway.* (From "La transformation des paysages congolais par le chemin de fer", H. Nicolai and J. Jacques, *Mémoires, Institut Royal Colonial Belge*, 1954.)

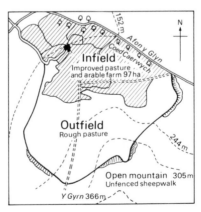

FIG. 6.—*Von Thünen's early model: Caerwych Farm, Talsarnau, North Wales.* (From *Geography Through Fieldwork: Book 2*, T. Bolton and P. A. Newbury.)

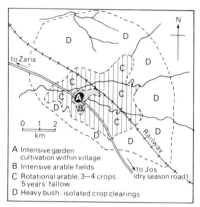

FIG. 7.—*Von Thünen's early model: Soba Village, Northern Nigeria.* (From "Land-use in Soba, Zaria Province, N. Nigeria", R. M. Prothero, *Economic Geography*, Vol. 33 (1957), pp. 72–86.)

Fig. 1(*b*)) whose related land-use zones interrupted those of the central city. He subsequently undertook various studies to assess the modifying effects on land-use of variations in such factors as soil fertility, climate and relief, and considered the implications of taxation, subsidies, import duties, etc.

Applications of von Thünen's models

The diverse applications of von Thünen illustrated in Figs. 2–10 embrace a wide range of geographical situations, historical contexts, and scales. While Figs. 2–5 demonstrate his later modified model, Figs. 6–10 give further examples of the early model of concentric land-use

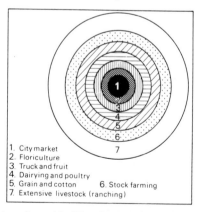

FIG. 8.—*Von Thünen's early model: Edwards Plateau, Texas, U.S.A.* (From "An economic study of a typical ranching area on the Edwards Plateau of Texas", Youngblood and Cox, Bulletin No. 295, Texas Agricultural Experimental Station (1922).)

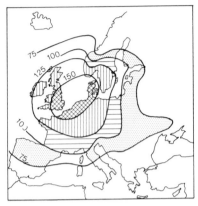

Fig. 9.—*Von Thünen's early model: intensity of agriculture in Europe.* Index of 100 based on average European yields per acre of wheat, rye, barley, oats, corn, potatoes, sugar-beet and hay. (From original map by S. Van Valkenburg and C. C. Held, *Europe*, Wiley, 2nd edition, 1952.)

zones. However, it should be stressed that these form only an arbitrary selection, and that for further information on these or any other aspect of von Thünen's ideas the student should refer to the relevant books listed in the bibliography.

Von Thünen and Christaller

Whereas Von Thünen was concerned with land-use distribution around the central market of a homogeneous agricultural region and sought to explain this in terms of economic rent, or locational rent, Walter Christaller, a fellow German, was more concerned with the distributional pattern and relative sizes of the market and service centres of a region. Both, however, based their models upon very similar postulates (*see* pp. 4 and 11–12). In other words, von Thünen regarded the central isolated city on a homogeneous plain as fixed, and proceeded to build up his concentric land-use zones around it, using such variables as production and transport costs and market prices. This analysis was inverted by Christaller, who viewed the homogeneous plain, the land-uses and the transport costs as fixed, and the locational pattern of the settlements as variable.

Walter Christaller, his Work and its Significance

Walter Christaller (1893–1969), a geographer, published his famous *Central Places in Southern Germany* in 1933. He is considered here before his teacher Alfred Weber, on the grounds that his work is more directly relevant to agricultural geography. However, he owed a great deal to both von Thünen and Weber.

FIG. 10.—*Von Thünen's zones reflected in marketing patterns on a world scale.* United Kingdom, 1960–2, imports of selected horticultural and dairy products. Oblique Azimuthal Projection showing distances correct from London. The zones are characterised by declining intensity of imports to the U.K. market with increasing distance from the U.K. For full details of the territories included in each of the four zones, and for the trade statistics upon which the zones have been calculated, see M. Chisholm, *Rural Settlement and Land Use.*

Central Place Theory

In the Central Place Theory, designed to explain the spatial distribution of settlements and their hinterlands and how this is likely to develop, he devised postulates which, as already noted, are reminiscent of those put forward by von Thünen. These, explicitly stated or implicitly assumed, include the following.

1. There is an unbounded homogeneous plain over which complete ease of movement is possible.
2. A single mode of transport is available throughout the plain, and transport costs in all parts are proportional to distance.

3. The density of population is the same in all parts of the plain; all consumers have the same purchasing power and require the same level of goods and services.
4. Every part of the plain is served by one or other "central place", that is, market, service and administrative centre, and no area is left unprovided for.
5. Consumers visit only the nearest central place that is capable of supplying a particular good or service. Since a lower-order centre provides fewer such services, this may necessitate a longer journey to a higher-order centre for some particular purpose.
6. Suppliers of goods or services seek to locate at a point of greatest accessibility in order to maximise their profits, but no excess profits are made by any supplier at any central place.

Christaller's theoretical argument runs as follows. Consumers will not go unnecessarily far to obtain a particular good or service lest their purchasing capacity is absorbed by increases in transport costs. The maximum distance purchasers will travel to buy that good is called its "range". Similarly the suppliers of any good or service need to serve a certain number of patrons, and since the population density throughout the plain is constant, this will determine a certain size of market area, or the threshold of that good.

An entrepreneur, supplying goods and services, will seek to isolate himself from other competing suppliers so as to have, within the range of the good or service he provides, as many potential customers as possible. He will seek to exceed the market threshold of his good if possible. Since to isolate himself further beyond a certain point from one competitor brings him nearer to others, the distribution of entrepreneurs supplying one particular good or service is likely to assume a triangular lattice pattern (*see* Fig. 11(*a*)).

In practice, the distinctive ranges and thresholds of the individual goods and services will tend to fall into several clearly defined groups. Thus entrepreneurs supplying goods and services with a very limited range and threshold would join up to form a low-order centre. Such a place would sell or provide only everyday goods or routine services, e.g. a village with a junior school, a grocery stores, perhaps a garage. On the other hand such amenities as a polytechnic or university, a large fashion store or a theatre, which are less frequently patronised and by fewer people out of the total population, are more likely to be established in the high street of a county town or in a large city, i.e. in a high-order centre. Such goods and services are described as low- and high-order respectively. Since each consumer will visit different levels of settlement more or less frequently for their distinctive order of goods or services, it follows that the large market area of a high-order centre will co-extend with a whole group of market areas belonging to low-order centres. Several such ranks of order develop.

How do these market areas develop? The maximum market area of any centre may be represented as a circle, the radius of which is the

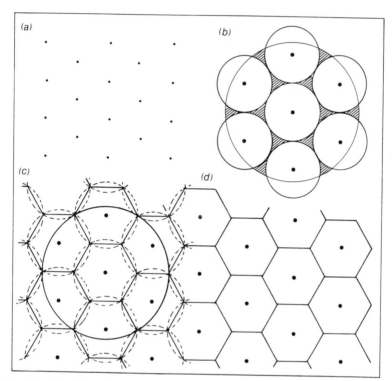

FIG. 11.—*How Christaller's "closely packed hexagons" evolve.* (*a*) Entrepreneurs seek to maximise market areas by isolating themselves, and in so doing assume a triangular "lattice pattern" of distribution. (*b*) When entrepreneurs avoid competition, consumers on margins are neglected in shaded areas of interstices. (*c*) When entrepreneurs extend market areas and compete, overlapping occurs. The consumers' efforts to minimise travelling expenses cause the overlapping portions to be bisected, and the market areas tend to assume a hexagonal shape as demonstrated in (*d*).

range of their goods and services. If the entrepreneurs do not vie with rivals in the adjacent market areas for custom, then such circles will barely touch, and this will mean that some customers in the interstices betweeen the circles will remain outside any market area. For all such customers to be reached, the circular market areas will be made to overlap. Then the natural consequence of customers seeking to minimise travelling expenses is for the overlapping area to be bisected, and so the circles are modified to form closely packed hexagons (*see* Fig. 11(*b*)–(*d*)).

Application of Christaller's theory to agriculture

Christaller went on to consider not only marketing but also traffic and administrative functions, using this same model. In the context of agriculture, however, we are concerned primarily with the first of these,

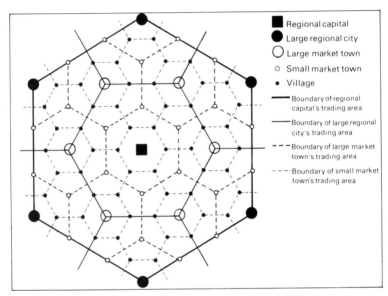

Regional capital
Large regional city
Large market town
Small market town
Village
Boundary of regional capital's trading area
Boundary of large regional city's trading area
Boundary of large market town's trading area
Boundary of small market town's trading area

FIG. 12.—*Christaller's primary lattice.* This diagram shows part of the regular hexagonal lattice which Christaller postulated to demonstrate his theoretical distribution and hierarchy of central places and their spheres of influence. The kind of area in which such a pattern of settlements might be expected to develop would be one in which the terrain, resources, population and purchasing power were evenly distributed, and in which movement was equally easy in all directions. (Reproduced from *A Geography of Settlements*, F. S. Hudson, Macdonald & Evans, 1977.)

and it is demonstrable that, given conditions approaching those of a homogeneous agricultural region, an even pattern of relatively equidistant market centres will emerge to provide the farmer with essential goods and services, including among the latter amenities for the sale of his agricultural produce.

Several echelons of settlement will develop (*see* Fig. 12). A distinctive feature of the lower order of market towns is the pannier market for selling dairy produce, vegetables, fruit and flowers at frequent intervals, while in the higher-order centres less frequent markets handle such commodities as livestock and grain.

In an old established agricultural region the lowest order of market has been declining or disappearing ever since the advent of modern specialised road transport, which, by reducing transport time and cost, has extended the range and market threshold of fresh agricultural produce.

The old range of low-order goods was related to the distance a laden man or beast could walk in an hour or so, always bearing in mind that

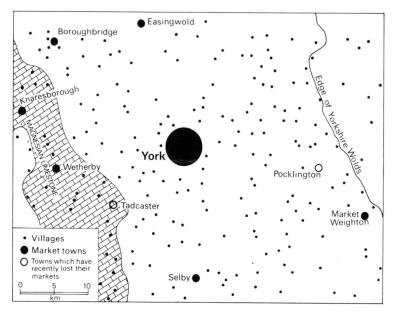

FIG. 13.—*Market towns and villages in part of the Vale of York.* York, the major route focus of the Vale of York, has become its dominant "central place". Around it is a scatter of villages and, at sparser intervals, small market towns. The latter have decreased in number as transport facilities have multiplied. There are a few villages on the thin chalky soils of the Yorkshire Wolds as the enclosure of these uplands for cultivation was begun little more than 150 years ago. (Reproduced from *A Geography of Settlements*, F. S. Hudson, Macdonald & Evans, 1977.)

a return journey must be made at evening after a heavy day. The physiological limits of man and beast of about 7 km per hour are reflected in many situations all over the world where markets in villages occur at something like 14 km intervals.

Modern transport has merely extended the market threshold of perishable, lower-order produce so that produce markets are only profitable if they are situated at intervals of 40–50 km from one another, despite attempts to combat this tendency by diversifying or opening at unorthodox times. While the lower order of markets has disappeared, higher orders of market continue to prosper in most county towns. A regional example to illustrate the operation of Christaller's Central Place Theory in an agricultural situation is given in Fig. 13.

Alfred Weber, his Work and its Significance

Alfred Weber, the German economist (1868–1958), first published his *Theory of the Location of Industries* in 1909, although it acquired greater popularity after translation into English in the U.S.A. some twenty years later, since when it has provided the basis of studies in industrial location. Whereas von Thünen had been concerned with land-use around a fixed market, Weber was concerned with the making of choices for least-cost, that is, most profitable, industrial locations. At first sight, such objectives appear wholly irrelevant to each other.

Weber's model was designed to assess the comparative production costs of industrial locations in relation to their transport costs, labour costs, and those economies to be derived from agglomeration. The model was based upon the following postulates.

1. There is a single isolated country, homogeneous in topography, climate, prevalent political and economic systems, race, culture and technical skills.
2. Entrepreneurs are profit-making and fully rational in seeking the most favourable locations.
3. Some natural resources, e.g. water, are ubiquitous (available everywhere), but many others, e.g. iron ore, are in fixed locales.
4. Labour is readily available but at fixed locations and at prescribed wage rates.
5. Consumer markets of prescribed sizes occur in certain fixed locations.
6. Raw materials and markets are available to all entrepreneurs in conditions of perfect competition.
7. Transport costs are directly proportional to weight and distance.

In his model, Weber's first concern was to locate those points where transport costs were minimal. He then sought to ascertain how cheaper labour, or economies available through agglomeration in certain locations, might draw production away from the location offering the lowest transport costs. Case A and Case B below illustrates the effects of transport costs on the location of industry as envisaged by Weber.

Case A: one market, one raw material source

In this situation there are three possible locations (*see* Fig. 14).

1. *Ubiquitous raw material.* Since raw materials are available anywhere at no transport costs, the factory will be located at the market.

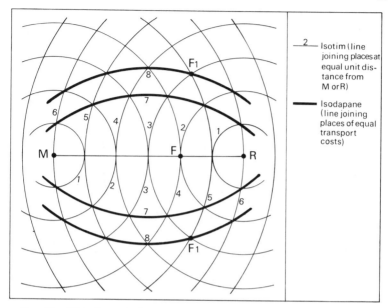

—2—	Isotim (line joining places at equal unit distance from M or R)
▬▬	Isodapane (line joining places of equal transport costs)

Fig. 14.—*Weber's simple Case A, and the added variable of lower labour costs.*

2. *Raw material at specific location, no weight loss in manufacture.*
In Fig. 14, the raw material is at R, which is six units of transport from the market (M). The cost of transporting the raw materials from R to M would be six units of transport. The cost of transporting the finished product from R to M would be six units of transport. The cost of transporting the raw material to a point on the line R–M, and then transporting the finished product to M, would also be six units of transport. The factory will therefore be located at R, at M, or at any intermediate point on the line R–M.

3. *Raw material at specific location, but weight loss in manufacture.*
In Fig. 14, the weight loss is assumed to be 50 per cent. If the factory is at M, then two units of raw material are required for one of product, so twelve units of transport will be required for each unit of product. If the factory is at R, two units of raw material are reduced to one of product, and only six units of transport are required to take one unit of product to the market (M). The factory will therefore be located at R.

Case B: one market, two raw material sources
In this situation there are four possible locations (*see* Fig. 15).

1. *Raw materials (R1 and R2) ubiquitous.* The factory will be situated at the market as in Case A.

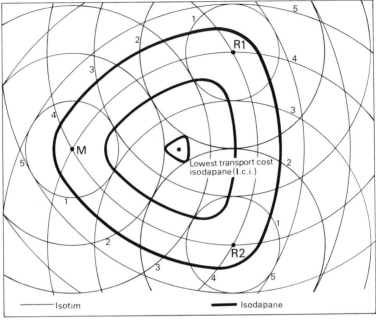

FIG. 15.—*Weber's more complex Case B.*

2. *R1 ubiquitous, R2 at specific location, neither has any weight loss in manufacture.* In this case transport is only required from R2 to market, where the R1 is readily available for manufacture. The factory will still therefore be located at the market.

3. *R1 and R2 both at specific locations, neither has weight loss in manufacture.* In Fig. 15, both R1 and R2 are four units of transport from M. If the factory is located at R1 or R2, it requires four units of transport to take the raw material from R1 to R2 or vice versa, and eight units to take the finished product (R1 + R2) to M, making a total of twelve units of transport. However, if the factory is at M, four units are required to bring R1 to M, and four units to bring R2 to M, making a total of eight transport units. If the factory is at l.c.i. (lowest transport cost isodapane), two units of transport are required to bring R1 to l.c.i., two units to bring R2 to l.c.i., and four units to take the finished product to M, again making a total of eight units of transport. The factory will therefore be located at M or l.c.i.

4. *R1 and R2 both at specific locations, both undergo weight loss in manufacture.* In Fig. 15, the weight loss is again assumed to be 50 per cent. If the factory is at R1 or R2, it will require four units

of transport to bring R1 to R2 or vice versa, and four units to take the reduced finished product to M, making a total of eight. If the factory is at M, it will also require eight units of transport, four to bring R1 and four to bring R2. However, if the factory is at l.c.i., two units of transport each are required to take R1 and R2 to l.c.i., and two units to take the reduced finished product to M, making a total of only six units of transport. The best solution therefore is to locate the factory at l.c.i., so that both raw materials are transported a minimum distance before manufacture reduces their bulk and the final product can travel to market more cheaply.

Lower labour costs

Weber noted that transport costs are, however, readily offset if wage rates are lower in an alternative location. The factory could profitably be placed on an isodapane of higher value provided the labour costs

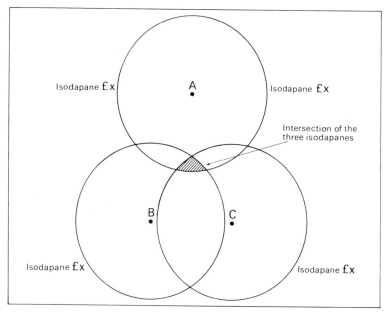

Fig. 16.—*Weber on the effects of agglomeration*. For three factories, the least transport cost locations are at A, B and C respectively. The isodapanes joining the points where the transport costs will be an additional £X are marked by the circles round A, B and C. Provided that the economies of agglomeration accruing to the three factories would be more than £X, it will be profitable for them to agglomerate within the area at the intersection of the isodapanes. (Based on diagram in *Human Geography: Theories and their Applications*, M. G. Bradford and W. A. Kent.)

at the new location were correspondingly lower. The deflection brought about by a lower labour-cost location is illustrated in Fig. 14. Production at factory site F1 will be cheaper than at F on line M–R, provided a saving derived from lower unit labour costs offsets higher transport costs on isodapane 8. At F, transport costs six units, labour four units and production therefore ten units. At F1 transport costs eight units, but if labour costs only one unit, production will cost nine units.

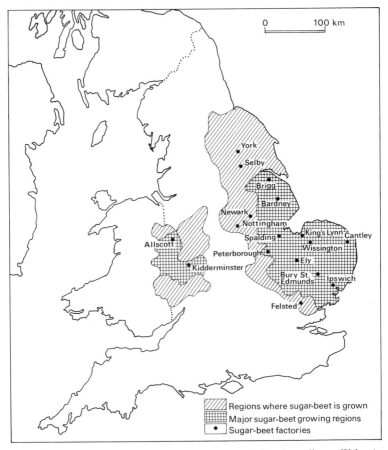

Fig. 17.—*Weber's weight-loss hypothesis tested against reality.* According to Weber (*see* Figs. 15 and 16), industries utilising raw materials which undergo substantial weight loss in manufacture are likely to be located nearer to their sources of material than to their markets. That this is borne out in the case of sugar-beet factories, whose final product, beet sugar, weighs only 12.5 per cent of its raw materials (coal, lime and sugar-beet), is demonstrated by their being located near to the sugar-beet production areas.

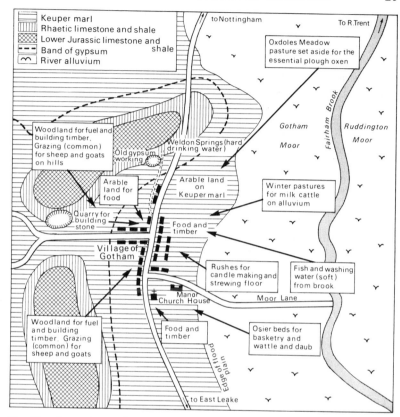

FIG. 18.—*Economic considerations which guided Anglo-Saxon settlers to choose Gotham, Nottinghamshire, for the site of a village.* (From *Geography Through Fieldwork: Book 2*, T. Bolton and P. A. Newbury.)

In this situation a new source of raw material might also be found which would then reduce the costs of transport at the new factory site, still further reducing the total production costs of the plant.

Effects of agglomeration

Weber also examined the effects upon industrial location of economies made possible by agglomeration (*see* Fig. 16). Thus firms sharing amenities available on a shared site can enjoy economies which, once again, might serve to offset higher transport costs on such a favoured location. Weber, it should also be noted, argued that such economies might also accrue to firms which increase their scale of production and achieve economies of scale, and argued that in some situations agglomeration may be a diseconomy if it causes a rise in land values, when deglomeration may offer the economies.

Relevance of Weber to agriculture

The relevance of Weber's model to agriculture may not be immediately apparent. However, as Bradford and Kent point out in *Human Geography: Theories and their Applications*, certain agricultural industries illustrate Weber's weight-loss hypothesis, as, for example, the sugar-beet factories of England and Wales seen in relation to the agricultural regions in which sugar-beet is an important crop. This kind of consideration is particularly relevant to the agricultural regions of East Anglia and the Fens, as regards both settlement and, through it, land-use (*see* Fig. 17).

Furthermore, as argued by Chisholm in his *Rural Settlement and Land Use*, there is no fundamental difference between an entrepreneur choosing an industrial location and an agricultural entrepreneur choosing the best location for a farm. This observation gives us an indication

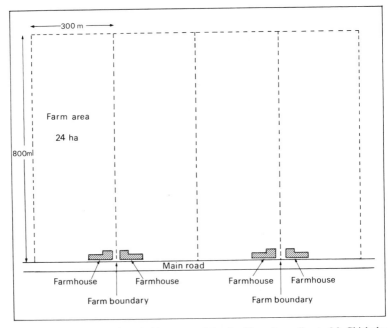

Fig. 19.—*Dutch planned farm-holdings on reclaimed polders.* According to M. Chisholm, *Rural Settlement and Land Use*, this pattern of holdings is a compromise, offering cheaper laying out and better social accessibility, but slightly higher operational costs than would square holdings with houses centrally placed. Most houses for farm labourers are provided in villages somewhat away from the farms.

of the kind of situation in which Weber is most relevant. Chisholm has cited the Anglo-Saxon colonists who settled all over England during a period from the fifth century onwards, when village lands were selected intelligently, if not scientifically, on the basis of economic considerations (*see* Fig. 18). The same kind of considerations will have influenced the early American colonists, while in the present century they are the dominant factor in the systematic planning of land-holdings for new farm-land created by reclamation in the Dutch polders and elsewhere (*see* Fig. 19).

A number of scholars have subsequently developed upon the work of von Thünen, Weber and Christaller, and the student should consult their works where possible. Notable among these have been the contributions of A. Lösch to settlement theory, and of W. Isard and E. M. Hoover to locational theory (*see* bibliography).

Chapter II

The Agricultural Systems of the World

What is meant by an Agricultural System

PETER HAGGETT, in his *Geography—a Modern Synthesis*, gives the following very generalised definition of a system as "a group of things or parts that work together through a regular set of relations" and further comments that "geographers are particularly interested in systems which link together man and environment". Agricultural systems, conforming as they do to this prerequisite of Haggett's, are supremely worthy of the geographer's interest, since they represent a primary aspect of man's response to his environment.

Like all living species man is motivated to strive towards individual and biological survival, and for these purposes he requires air, drink, food and shelter. From earliest times, in search of these things, man has acquired the fundamental skills that constitute the basis of agricultural technology.

An agricultural system is made up, in Haggett's words, of "a group of things or parts that work together through a regular set of relations". Any particular system is made up of the sum total of these, and will differ from other systems not only in the character and variety of these "things and parts", but also in the complexity of their interrelationships.

The character of each system has been moulded, as suggested on p. 31, by the distinctive kind of physical environment in which it has evolved, but it will also reflect the cultural history of the people who have developed or adapted it. The more technically advanced systems have acquired a degree of independence from the limitations imposed by the natural environment which the more primitive systems could not enjoy. Consequently there are examples of such systems being applied to hard and poorly endowed environments while more favoured environments remain in the thrall of primitive peoples. In general, however, there is a tendency for the more advanced systems to oust the less advanced from all but marginally productive regions.

It would be misleading to see any system simply as a response to some kind of environment, for although it is true that the constituent "things and parts" and "regular set of relations" have usually been acquired in a first situation as a response to the sum total of geographical conditions there, yet it is also true that the more successful systems may be capable of survival when transferred to other similar, or on occasions quite dissimilar, environments by the people who originally developed it, or by other people who have adopted it. Thus it was that the technically advanced Roman conquerors of Britain were able to introduce viticulture into southern England in face of an inimical natural environment—a system wholly out of keeping with the more predictable agricultural responses of the indigenous Britons. This is not, of course, to understate the wholly independent flowerings of the same agricultural system in different continents through parallel development in response to similar environments. This must of necessity be the case when considering the primitive systems evolved by subsistence peoples wholly incapable of making commercial contact with other regions in which the same system has evolved coincidentally.

In any of these several ways, then, a system may come to exist in several locations around the world. Within any one such location there will exist a number of similar economic units of agricultural production, e.g. farms or ranches. In each such unit, as in any of man's economic activities, some practical combination of land, labour and capital (the factors of production) will be directed towards the provision of selected agricultural commodities. To reach an understanding of a particular system it is convenient to apply the economic classification of land, labour and capital to Haggett's "things or parts".

As regards agriculture, "land" is represented literally by an area of ground upon which some characteristic pattern of land-holding, distinctive in size, shape, internal arrangement and degree of fragmentation or consolidation, is likely to recur.

"Labour" is applied to land in certain organised relationships and employing operational skills in accordance with the regular and effective pattern which accords to that system, although modified by local differences in the culture and technological attainment of the particular people concerned.

"Capital" for the more primitive systems will be confined to "working capital" afforded by a few simple implements or the seed and livestock employed in production. It is in the greater application of capital that the most significant difference can be seen between the primitive and the more advanced agricultural system. In the latter increasing levels of capital are employed to provide any drainage works and irrigation schemes, purpose-built buildings and sophisticated machinery. In addition, more production inputs, such as fencing, fertiliser or machine fuel, are employed. Where there is a regular production surplus the unit may also require and contribute towards the provision and upkeep of shared communications, transport and marketing facilities. The relatively small optimum scale of single productive units in agriculture is rarely large enough to warrant the provision of these facilities by the single unit.

The co-ordination and organisation of all these "things or parts" within one productive unit going to make up an agricultural region operating under some particular system requires what Haggett called the "regular set of relations" to give them coherence and cohesion. Primitive systems evolved their relations empirically, but modern systems are increasingly called upon to devise some well-defined rationalised structure, in order the more effectively to incorporate technological discoveries designed to promote greater competitiveness and achieve fuller independence from the forces of nature.

In all systems, whether simple and primitive or complex and sophisticated, corresponding "things or parts" will be discernible, organised into similar "regular sets of relations". Any system may therefore be described in terms of its land-holding structure, its land-use, its organisation and mode of application with regard to labour, the operational skills and techniques employed, and all the ways in which capital is employed in terms of plant, machinery, technical aids, seed and livestock, etc. Where regular surpluses are produced, it is also relevant to describe the transport, marketing and ultimate consumers of these, since the pattern of marketing may be the most powerful determinant of all in the subsequent evolution of the more advanced, highly capitalised and market-orientated systems.

The world systems which have been identified in this book include ones represented in every continent of the world, and ones operated by people at every level of technology and civilisation. An attempt has been made to place them, somewhat arbitrarily, into an order of evolution, and this evolutionary process is demonstrated where it is relevant to the origins of a particular regional example, or when one system is seen to be in transition towards another.

Systems may come and go in any given regional situation, but this does not of itself diminish the value of learning about an agricultural system and its corresponding way of life simply because it is in decline or has even disappeared, since such systems and their development may be paralleled in other places and at different times.

The World Distribution of the Main Systems

It will be apparent from Fig. 20 that the most primitive agricultural systems, once far more widespread, are being ousted out of all but the least productive, marginal lands. These are usually to be found in the extremely high, polar , latitudes or the low, tropical, latitudes, or else they are situated in the hearts of vast continental interiors, all places where almost unmanageable positive and negative extremes of temperature or precipitation may be encountered. These are subsequently described in Chapters VI and VII.

The need to give priority to the consideration of agricultural systems with greatest world economic significance has regrettably required me to consider only briefly several systems of very real academic interest

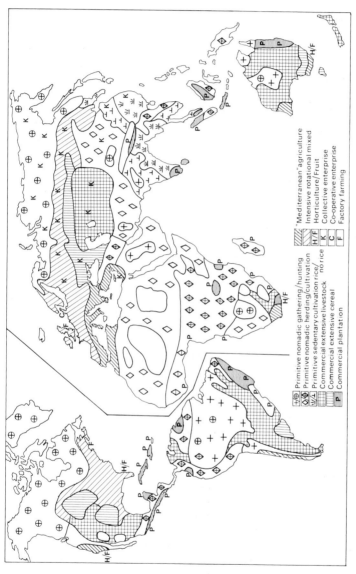

FIG. 20.— *The major agricultural systems of the world.*

from the point of view not only of the anthropologist but also of the historical geographer concerned with the causes of and processes underlying geographical change.

The primitive nomadic gatherers who inhabit regions with all-round growing seasons, or their more widespread counterparts the primitive nomadic hunters who range over some of the most inhospitable landscapes of the world, are not, in point of fact, true agriculturists at all. Yet they are pre-farmers, since man had to live before he was able to master the arts of cultivation and herding—in much the same way as Alexander Selkirk (the real Robinson Crusoe) was obliged to do, despite the heritage of civilisation he brought with him, when first confronted with the new "island" situation.

It will be shown that the primitive nomadic hunter can and has evolved to nomadic herding and thence towards commercial rancher. Similarly the nomadic cultivator will, under pressure of land hunger, adopt the semi-nomadic life of the bush fallower, and, where economic conditions permit, will eventually evolve towards the sedentary rotational system with fallow period, known best perhaps in the open-field system of the Anglo-Saxons, but practised today by some West African tribes. The intensive peasant subsistence farming system still practised in large areas of South-east Asia may be looked upon as a rough parallel to medieval Europe's open fields.

The commercial extensive systems described in Chapters VIII and IX grew up in response to pressures of rapid commercial and industrial expansion in Western Europe, obliging emigrants to settle in remote underpeopled continents, where they soon found it profitable to adopt labour-intensive methods of farming. These made it possible for them to produce vast surpluses to export for the support of the rapidly expanding industrial populations of European countries they had left behind.

Plantation agriculture, considered in Chapter X, developed in a similar situation to the commercial extensive systems, for it grew out of the work of a small number of people of European origin who sought to produce profitable surpluses of much-needed tropical produce for export to Europe. However, whereas the great commercial livestock and cereal regions were in latitudes where the climates, though severe, were tolerable to Europeans, most plantations were developed in regions where, according to the existing level of medical knowledge, conditions were barely, if at all, suitable for Europeans. The owners of plantations, too few or unable, therefore, to undertake the work of cultivating their holdings themselves, came to depend on the exploitation either of the local indigenous population or of imported slave or indentured labour from other tropical areas. More recent emancipation of tropical peoples has disadvantaged plantation agriculture, bringing about considerable modification of the system.

The mediterranean system and the intensive rotational mixed system are practised by some of the most affluent agriculturists in the world, enjoying as they do close proximity to large industrial urban markets. Their exponents in the Southern Hemisphere, notably in Australia and South Africa, are farmers of European ethnic origins, who up to present times have regarded European countries as their "natural" trading partners. These systems are described in Chapters XI and XII.

Of all agricultural systems, that which goes under such names as market gardening or fruit growing, truck farming or horticulture is one of the most specialised, and is important far beyond the small area of land devoted to it. It is specialised as regards the small select localities which are often accorded to it, close to dense urban markets or to food-processing facilities with which the system is often in symbiotic association, or on rapid communication lines to one or other of these. It is also specialised as regards the particular qualities of soil or climate which make some small locality specially reserved for the production of one particular fruit, vegetable or flower. The letters H/F on Fig. 20 are intended to denote such small specialised regions, too small to be indicated by shading on the map. This expanding system is considered in Chapter XIII.

A quite different distinction, related less to regional physical differences than to divergent political and economic ideologies, lies behind the two contrasted systems, collective and co-operative farming, considered in Chapters XIV and XV and indicated by K and C on Fig. 20. (Letters are used again because the systems co-exist with such diverse natural agricultural regions.)

It will be observed that Marxist Communism, which originated in nineteenth-century Germany, is still, as a governing politico-economic structure at any rate, almost exclusively confined to Eastern Europe and Asia. Capitalism, its much older protagonist, is largely confined to Western Europe, North America and those regions inhabited in the main by people of European ethnic descent, for example Australia. The "Third World", made up of the poor and underprivileged peoples of Africa, South America and the Indian subcontinent, stands for the moment in a state of non-commital bemusement, attracted equally by the blandishments proferred it by the former, and by the affluence half-promised by the latter.

It is significant that the collective and co-operative agricultural systems are largely confined to Eurasia. The collective is usually the Communist solution, although Israel, an anachronistic East/West state in a hectic "front-line" situation and with a corresponding yen for social stability, is aligned with the capitalistic West but has adopted the collective farm, or Kibbutz, for one element of its social structure.

The co-operative has found its strongest embodiment in Denmark, where it has enabled the small farm units of a tiny country heavily dependent upon agriculture to achieve a remarkable level of prosperity, but has been only restrainedly experimented with in most of the countries of Europe, and has had very little impact upon agriculture in the North American continent.

Finally, Chapter XVI discusses a system, factory farming, which is still only in its infancy. However, although almost wholly confined to two regions, it is an augury for the future. There is a real danger that in the next half century or sooner great megalopolises, sprawling conurbations many times larger than any that at present exist, may engulf the north-east quadrant and Pacific Coast regions of the U.S.A., Southeast England, and the area which includes the lower Rhine from the Ruhr to Rotterdam-Europoort, which has sometimes been called Europe's "Golden Triangle". If so, this system, born out of the economic necessity to provide abundant cheaper food for teaming millions of urban dwellers with little or no available agricultural land, may well become widespread. Other possible regions might include Tokyo, Calcutta and Bombay. Existing locations are indicated in Fig. 20 by the letter F.

In the next chapter consideration is given to the physical factors which helped determine the present world distribution of the major agricultural systems.

Chapter III

Physical Influences on Distribution of Agricultural Systems

THIS chapter is concerned with the physical factors that influence the world distribution of the main agricultural systems. It is convenient to consider these factors at three levels: first by describing the obstacles that have limited agricultural land-use to only a fraction of the earth's surface; secondly by examination of the factors that have operated to restrain or constrain agriculture at the continental level, for which purpose North America will provide the basis of study; and finally by consideration of the modifying factors that operate at the local level, illustrated by the Great Valley of California in particular.

Land available to Agriculture at the World Level

The physical obstacles to agricultural production are best appreciated if the student consults any good, advanced atlas, where they will be illustrated more fully and effectively than is practical here. The maps that should be examined include particularly those illustrating such relevant features as geology and structure, relief and drainage, ocean currents, pressure and wind systems, and, following on from these, those showing the world distribution of climatic types, soils and natural vegetation.

Reference to Fig. 20 will confirm that of all the above features, the most influential upon soils, natural vegetation and agricultural land-use at the world level have been the relative positions of the great continental land masses and oceans, climate (particularly through the influence of latitude in causing polar wastes and hot or cold deserts), and the occurrence of major mountain systems, which not only give rise to unproductive highland, but also act as climatic barriers at the world scale.

What is significant to the student at this level is that the total land area available to agriculture throughout the world is much diminished by the collective operation of such agencies. To summarise, of the world's total surface area, 75 per cent is covered by bodies of water, while the 25 per cent that remains is the total land surface. Of this total land surface:

1. *35 per cent is barren*, and made up of:

	Percentage total land surface	*Percentage total world surface*
polar cap and mountain	16	4
hot and cold desert	19	4.75

2. *65 per cent is potentially productive,* and made up of:

	Percentage total land surface	Percentage total world surface
forest, woodland and open range	36	9
pasture and meadow	19	4.75
arable land	10	2.5

Modifying influences at continental or local levels preclude effective agricultural use, partially or completely, over much of the potentially productive land area.

The above estimate is based on statistics from various sources brought together by Haggett in *Geography—a Modern Synthesis.*

Factors influencing Agriculture at the Continental Level

Probably the most significant influences at the continental level, operating in conjunction with one another, are latitude, altitude and degree of continentality. In addition, however, there are modifying factors such as large-scale relief features, sea currents adjacent to coastal margins, and erosional and depositional features of Quaternary glaciation, all of which exert far-reaching effects upon the character of the continental land mass concerned. Consideration of North America will help to illustrate the operation of physical factors at this level.

The effects of latitude on climate

The latitudinal position of a place has a direct bearing upon its prevailing temperatures. In general, low latitudes experience the maximum heat- and light-giving propensities of the sun, and high latitudes the minimum. This, however, is certainly not because the poles are further from the sun, because an additional 6,400 km is hardly significant when the sun's rays have already traversed some 150 million km before they reach the nearest part of the earth's surface.

The earth's troposphere extends about 9 km above the poles and 17 km above the equator, these heights varying with the seasons. The water vapour in the troposphere absorbs the sun's heat and light, and the more directly the sun's rays penetrate the troposphere, the less they are filtered by it. In addition, the more vertically the sun's rays strike the surface of the earth, the smaller the area with which they come into contact, and the brighter and hotter they will therefore make it.

These effects are illustrated in Fig. 21. The figure shows three rays of the sun, equal in heat and light intensity and approximately parallel to each other. Ray A, at the equator, penetrates the atmosphere and reaches the earth directly, with least heat loss and maximum intensity of rays. Ray B, at the mid-latitudes, penetrates the atmosphere and

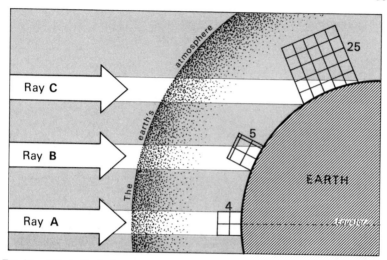

FIG. 21.—*Effect of latitude on temperature*. This diagram illustrates the conditions pertaining at the equinoxes (21st March and 23rd September).

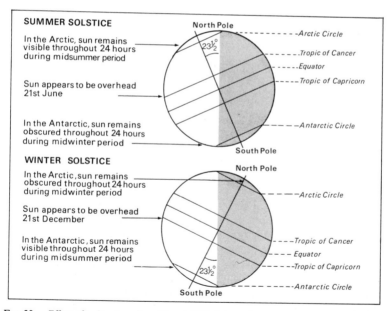

FIG. 22.—*Effect of inclination of earth's axis*. This brings about seasonal climatic changes, and complicates the effects of latitude on climate except at the equinoxes, when the effects of inclination are momentarily nullified, because the sun's rays strike the earth vertically.

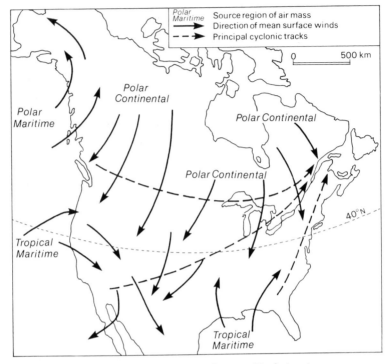

FIG. 23.—*January climatic conditions in North America.*

reaches the earth obliquely, with greater heat loss and rather less intensity of rays. Ray C, at the poles, penetrates the atmosphere and
reaches the earth very obliquely, with maximum heat loss and minimum
intensity of rays. The figure also illustrates the variation in land area
heated by each ray: 4 units at the equator, 5 at the mid-latitudes and
25 at the poles.

There is an additional complication to be taken into account since
it directly affects the climate of regions in all latitudes between 50° N.
and 50° S. This is the inclination or tilt of the earth's axis of rotation
at an angle of $23\frac{1}{2}°$ out of the vertical, as illustrated in Fig. 22. As the
earth follows an orbit around the sun once every $365\frac{1}{4}$ days, rotating
every 24 hours as it goes, so it inclines towards the sun, first exposing
the Northern Hemisphere most fully at its summer solstice (21st June),
and then exposing the Southern Hemisphere most fully during *its* summer solstice (the Northern Hemisphere's winter solstice, 21st
December). At two other dates only, 21st March and 23rd September—
known as the equinoxes—the tilt towards the sun is briefly non-existent,

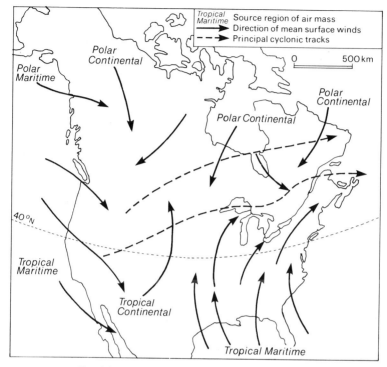

FIG. 24.—*July climatic conditions in North America.*

and all latitudes enjoy one night (and day) of equal duration, hence the name "equinox", from the Latin words for "equal night".

The significance of latitude in this context is therefore complicated by the seasonal variations in insolation that result. Not only do low latitudes in each tropic receive more insolation than the high latitudes, but the zone experiencing maximum insolation at any given time migrates from $23\frac{1}{2}°$ N. to $23\frac{1}{2}°$ S. over a period of half a year, and then back again.

All the corresponding latitudinally disposed pressure zones, and their related wind systems, therefore shift north and south over a rather smaller amplitude. As a result, seasonal variations in wind directions and precipitation as well as temperatures occur, brought about by the earth's inclination.

This seasonal swing of the pressure and wind systems affects all the U.S.A. in that, whereas tropical winds are dominant only over the Gulf Region in January, by July these have extended their influence north of the Great Lakes into southern Canada (*see* Figs. 23 and 24). This distribution is also brought about in part, however, by a massive seasonal high-pressure air mass over the broad, ice-bound continental extent of northern Canada in winter, one which is dissipated during the following hot summer season. As will be seen later, the effects of decreasing poleward insolation are reflected most forcibly in the natural vegetation and agricultural land-use zones of North America, although latitude is only one of several significant influences to be considered.

The effects of altitude upon climate

Nowhere in the U.S.A. are the effects of altitude more graphically demonstrated than in the eastern part of the Colorado Plateau, where

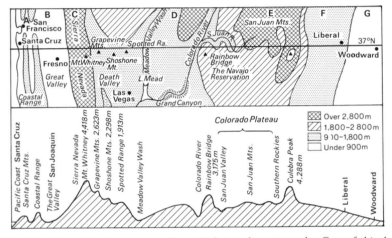

FIG. 25.—*Map and section across south-west U.S.A. to demonstrate the effects of altitude on climate and land-use and of a mountain barrier on rainfall distribution.*

the Navajo Indians eke out a miserable living in one of the most climatically extreme regions in the world (*see* Fig. 25).

The effects of altitude on climate, natural vegetation, and agriculture in the Navajo Territory can be summarised as follows:

> below 1,500 m: dry, hot; desert cacti; maize and potatoes by irrigation
> 1,500–1,800 m: cooler; grass, sage-bush; winter sheep pasture
> 1,800–2,100 m: cooler; shrubs, trees, juniper and piñon
> above 2,100 m: well-watered highlands; yellow pine and grassland; summer sheep pasture

Originally hunters and gatherers, the Navajos were introduced to sheep and goat herding by the Spaniards some 400 years ago. Today many of them make a living by nomadic herding. After the spring thaw they take their herds up to the summer pastures at some 2,000 m where the sparse rainfall is marginally higher and more effective at the prevailing lower temperatures. This allows the lower plateau level at about 1,500 m to recover from the heavy grazing of the previous winter. In autumn they return to these lower levels. By means of this seasonal migration the Navajos are able to avoid the peak mean summer temperature in July of about 24°C, and their flocks can graze on the moister mountain pastures where the temperatures are averaging about 20°C over the same period. Other Navajos remain in the lower areas, growing, in particular, crops of maize and potatoes employing simple irrigational devices.

The effects of distance from sea upon temperature

It is readily observed by any seaside visitor that whereas the sandy beach rapidly loses its heat as the sun goes down in the evening, the sea may still remain warm and inviting after dark. Similarly the sea is never so unenduringly hot as the beach may become at midday. These differen-

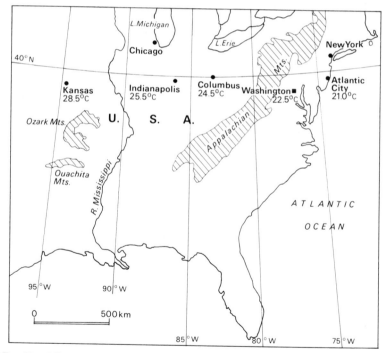

FIG. 26.—*Effect of distance from sea on the annual temperature range of selected stations in the U.S.A.*

tial heating and cooling rates displayed between beach and sea have their counterpart in the seasonal pattern of heating and cooling which takes place between the large continental land masses and oceans. Because, moreover, air masses over land and sea correspond in temperature to the surface they overlie, and because there is an interchange of land and sea breezes in coastal regions, the sea's greater warmth in winter, or coolness in the height of summer, may be reflected in these same coastal regions, which experience what is described as a maritime climate. The effects of the sea decline inland, so that we may describe conditions inland as showing increasing continentality. Maritime climates therefore tend to display milder winters and cooler summers, which culminate in a small annual temperature range. They also tend to display a heavy rainfall. Further inland, with increasing continentality, winters are more severe and summers hotter, and these extremes are reflected in a large annual range of temperature.

Figure 26 shows that the annual temperature range of selected stations in the U.S.A. is increased by 7.5°C over a distance inland of about 1,500 km. It should be mentioned that the more inland of the

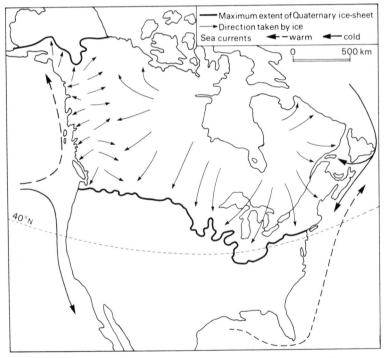

FIG. 27.—*Fullest extent of glaciation in North America, and the warm and cold sea currents along its coasts.*

stations indicated in the figure are at a greater altitude than those near the coast. Since we are concerned here with temperature ranges, and since the greater altitude is a constant factor in lowering the temperatures experienced by those stations at all seasons, it does not materially affect this illustration, but clearly the greater severity of winters in the Midwest are a function of this greater altitude as well as greater continentality.

A warm sea current along the coast may serve to reinforce the generally mildening influence of the sea upon the climate of coastal regions by warming the air mass over the sea at all seasons, a process particularly significant in its effect in raising winter temperatures within the maritime areas. The Gulf Stream (*see* Fig. 27) affects the south-east coast of the U.S.A., but its influence, although much diffused, is also felt when, as the North Atlantic Drift, it passes up the coast of Western Europe and its ameliorating effects are even experienced in Arctic waters. By the same token, the cold Labrador Current intensifies the severity of Newfoundland's climate.

The modifying effect of a mountain barrier upon rainfall

Figure 25 has already been referred to in the context of the effects of altitude upon temperature. This, however, is only one side of the picture. Not only are maritime air masses, heavily laden with water vapour, cooled both by their ascent and by contact with cold mountain flanks, and so obliged to release their rain; but also the same air masses, warmed by their descent on the lee side of the mountains, become "drying" winds and result in rain shadow or partial rain shadow conditions in the continental interiors beyond the mountains.

The heavy orographic (relief) rain experienced on the Coastal Ranges, Sierra Nevada and Southern Rockies contrasts strongly with the arid conditions displayed by the Great Valley (where agriculture is almost wholly dependent upon irrigation), and in intermontane plateau regions such as Colorado, or on the semi-arid Western Plains beyond the Rockies. The differences in rainfall in these regions can be summarised as follows:

Region (see Fig. 25)	Average rainfall per annum (mm)
A	560
B	130—250
C	1,000+
D	250–500
E	760–1,000
F	250–380
G	630

Effects of glaciation

The effects of glacial erosion and deposition, or the absence of glaciation, have also contributed to the agricultural land-use contrasts evident in North America (*see* Fig. 27). In the bare and eroded expanse of the Laurentian Shield of Canada (taiga and tundra in Fig. 28) there is an almost total lack of soil over large areas; what does exist is podsolic, and no agricultural land-use is indicated in Fig. 29. In contrast, the mature czernozems indicated further south on the prairies, and given over to arable farming, occur in a region unaffected by the Quaternary glaciation. Had this soil been removed, even the conducive climate of the Midwest could not have made up for it.

Conclusion

Russian and American geographers and pedologists, who have been accustomed to investigations conducted on the continental scale, have emphasised that latitude, and subsequently climate, have been the dominant factors in the distribution of soils at the continental level,

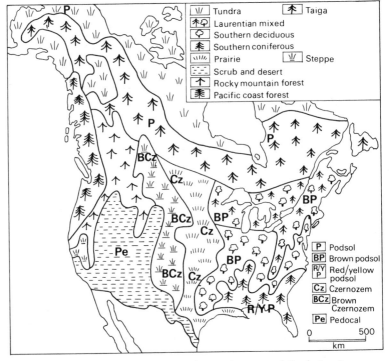

FIG. 28.—*Climate, soil and natural vegetation zones of North America.*

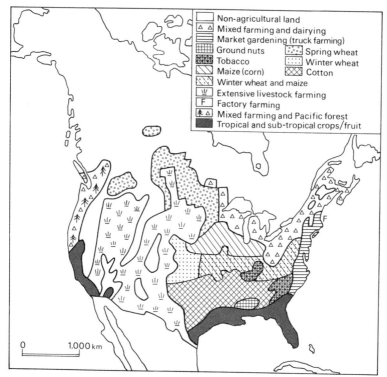

Fig. 29.—*Agricultural crop zones of North America.* These can readily be fitted into the more generalised scheme of agricultural systems of the world.

while climate and soil together exert an influence upon the natural vegetation and land-use zones, and bring about the familiar latitudinal alignment shown in Figs. 28 and 29.

Notable exceptions to this are the zones adjacent to both littorals. The Western Cordillera and the Appalachians have, by raising altitude, led to decreases in temperature. Further, the Western Cordillera has imposed more or less severe rain-shadow conditions over much of the western half of North America and brought about the arid marginal conditions which have prevented arable farming from ousting extensive livestock farming out of the Great Plains as has happened in the Midwest.

Lack of closer correspondence between the natural vegetation and agricultural land-use zones is in part a result of human factors which are considered separately in the subsequent chapter.

Factors influencing Distribution of Systems at the Local Level

It remains now to consider factors which at the world or continental levels have exercised only a minor role, but which are predominant at the local level. As already observed, their environment has led American and Russian geographers and pedologists to lay stress upon latitude and climate as the major determinants of soils and land-use zones. In the early days of their investigations, the smaller environment of British geographers led them to stress geology, structure and relief as the major determinants, and so, at the more intimate local scale, they are.

Geology determines the physical and chemical composition of the "parent materials" from which a soil is derived. At the continental scale this factor is overridden by the climatic influences which regulate the soil's subsequent development, but this effect cannot be seen so clearly over a small area.

Structure determines differences in relief and drainage which affect subsequent soil development and land-use.

To examine these factors more closely, we will consider briefly one small region selected from the North American continent, namely the Great Valley of California in its mountain frame, occupying the extreme south-west of the U.S.A.

The Great Valley of California

In Fig. 30 the three main physical divisions of this extreme south-west sector of U.S.A. are shown. Below are described the important factors of geology, structure, relief, climate and natural vegetation which help to explain the contrasted agricultural land-uses which occur in so small an area.

1. *The Coastal Ranges.*
   (*a*) Geology, structure and relief: folded, faulted ranges, 760–1,520 m.
   (*b*) Climate:

	Jan. mean	July mean	Rainfall per annum
Eureka	—	13.5°C	940 mm
San Francisco	10°C	14.0°C	570 mm
San Luis Obispo	—	17.5°C	560 mm

   (*c*) Natural vegetation: forest in north, evergreen chaparral in south.
   (*d*) Agriculture: dairying in northern valleys; fruit, vegetables and flowers in Santa Clara Valley; soft fruit and cattle in Salinas Valley.

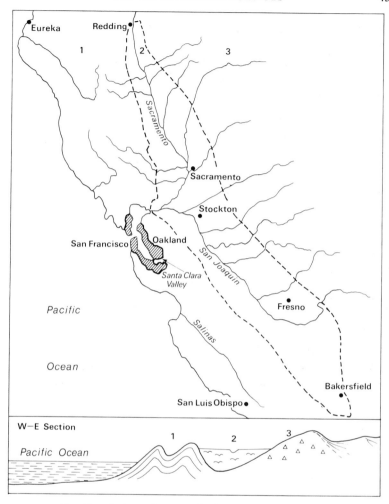

Fig. 30.—*Regions adjacent to the Great Valley of California.*

2. The Great Valley.

(a) Geology, structure and relief: synclinal structure, 640 km from north to south, and 80 km wide; floored by alluvium carried down from adjacent Sierras.

(b) Climate:

	Jan. mean	July mean	Rainfall per annum
Redding	—	27.5°C	940 mm
Sacramento	7.5°C	22.5°C	490 mm

	Jan. mean	*July mean*	*Rainfall per annum*
Stockton	—	23.0°C	360 mm
Fresno	7.0°C	27.0°C	180 mm
Bakersfield	—	28.0°C	150 mm

- (c) Natural vegetation: steppe in north, sage-brush in south.
- (d) Agriculture: grapes, peaches, apricots, oranges, nuts, sugar-beet, vegetables, cotton under irrigation; cattle on drier margins.

3. *The Sierra Nevada.*
- (a) Geology, structure and relief: massive block of volcanic, metamorphic and sedimentary origins, uplifted and tilted with steep side inland. Summits average 4,250 m.
- (b) Climate: because of altitude, cold all year; rainfall unpredictable, but averaging 1,000 mm.
- (c) Natural vegetation: coniferous forest, Douglas fir, redwood, pine, etc.
- (d) Agriculture: summer pasture for cattle and sheep from Great Valley in lower altitudes; some forestry.

It will be observed that the Great Valley is a fertile alluvial area situated in a syncline between two upstanding mountain ranges, and built up by the rivers Sacremento and San Joaquin. It will be noted too that temperatures decrease northwards along both the Coastal Ranges and the Valley. Temperatures, however, are more extreme in the Great Valley than on the coast, because the interior is isolated from the maritime influences by the Coastal Ranges. San Francisco, for example, has a climatic range of only 4°C, whereas Sacramento has a range of 15°C.

Comparisons between the rainfall figures indicate that, whereas Redding at the north end of the Great Valley has the same annual rainfall as adjacent Eureka on the coast, the rainfall in the Great Valley decreases far more rapidly southwards than does that of the coastal plain, e.g. Bakersfield gets only 150 mm of rainfall per annum against San Luis Obispo which gets 560 mm per annum. The hot, dry summers of the Great Valley reduce the surviving natural vegetation to steppe (*see* Fig. 28) and make agriculture wholly dependent upon irrigation from the snow-fed waters of the Sacramento and San Joaquin.

The natural vegetation generally reveals increased aridity from north to south, and in particular reflects the aridity of the Great Valley. With irrigation, however, to overcome this handicap, the essential fertility of the Great Valley is demonstrated in the profusion of agricultural products, of which the list above is at best a summary. Altitude, of course, as well as thinner soils except in the Santa Clara and Salinas Valleys of the Coastal Ranges, restrains agricultural activities in both this area and the Sierra Nevada Mountains to the east of the Great Valley.

To conclude, this small area demonstrates the considerable modifying influences upon agriculture exerted locally by geology, structure, relief, and thereby climate and soils. In this case the contrasts are sufficient to create a local pocket of intense agricultural activity, similar to the Rhine Rift Valley in Europe. The influence of such local factors is far more widespread, however, than this would suggest, usually modifying agricultural systems rather than, as in this case, delimiting them.

Chapter IV

Human Influences on Distribution of Agricultural Systems

Introduction: Technology v. Geographical Determinism

A T an early stage of his development, man, in his efforts to satisfy basic needs such as food, drink and shelter, stumbled upon the first rudimentary arts of agriculture. In striving towards greater security and comfort, man evolved various patterns of social organisation to which we give the name civilisation. Agriculture has been an integral part of all civilisations.

Technology, which was made possible by civilisation, has been reflected in more effective and sophisticated systems of agriculture. Each new level of civilisation achieved has had the effect of changing man's ecological status, thereby making possible the support of a larger population. Each rise in population has in turn stimulated new technology, in this context new systems of agricultural technology and organisation. In modern times technology has brought about the advanced industrial society which has supplanted agriculture as the direct source of livelihood for the majority of people, and as the direct sustainer of civilisation.

Nevertheless, as long as man must eat and drink, agriculture will remain at the root of his survival, even though he is no longer as aware of the umbilical cord binding him to the earth. Man is still ultimately dependent upon agriculture, and, to that degree, upon the environment in which he practises agriculture.

The theory of geographical determinism, first promulgated by the French philosopher Montesquieu in the eighteenth century, argues that man—as a physical being with distinctive bodily attributes; as a thinking being, with thought patterns, communication skills and a philosophical inheritance; as an economic being, with agricultural, manufacturing and commercial technologies; and as a social being, with his socio-political institutions—is a product of his geographical environment, be it narrow and restrictive or varied and extensive.

Early man was forced to adapt to his environment in order to survive, and many early agricultural developments were just one aspect of that process of adaptation. As his technology has developed, however, man has been able to acquire some degree of independence from his environment, and so the influence of geographical determinism has softened into a possibilism, as man has wrested a degree of choice over his destiny from nature. This greater flexibility of human adaptations to environment is well illustrated in a consideration of world agricultural systems.

In this chapter it is intended to consider the origins of man; how at an early stage he acquired not only divergent physiological traits but also differences in technology and social organisation; how he has spread or transferred these characteristics by migrations or commercial associations over vast areas; how in some instances the diffusion of these has been inhibited by natural geographical barriers; how man's civilisation, including his agricultural systems, has evolved as he has striven towards greater ascendancy over his environment; and how, in short, we have arrived at the existing broad distributional patterns of mankind and of human civilisation. We shall then have gone a long way towards a fuller appreciation of the human factors which have influenced the existing world distribution of agricultural systems.

The Origins of Man and his Early Migrations

Our species, *Homo sapiens*, was derived from the anthropoid suborder of primates. Among these anthropoids divergent forms, including the anthropoid ape and the hominids, emerged. These hominids, identified by such fossil remains as Java and Peking man, are believed to have lived about 1 million years ago. From them more man-like forms—*Homo* though not *Homo sapiens*—were to develop, creatures such as Neanderthal and Rhodesian man. Finally, *Homo sapiens* emerged, at first contemporaneously with the hominids. Among these, for example, was Kenya man, a less specialised and therefore more adaptable creature than the hominids, which helps to account for his survival in face of environmental changes which were soon to sweep the hominids into oblivion.

Even in the short period of earth history since man first appeared,

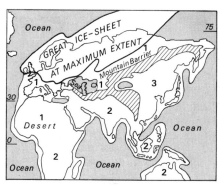

FIG. 31.—*Basic environments of* Homo sapiens *during the era of Quaternary glaciation,* c. *750,000–30,000* B.C.

the world has been subject to far-reaching changes as regards both its distribution of land masses and its climatic zones. Anthropologists believe they have identified three early natural environments (*see* Fig. 31) which correspond with the three primary races of man, from which subsequently several sub-races of man have also developed (*see* Fig. 32).

These three natural environments have each impressed, by a process of natural selection, a distinctive and durable set of physiological, or racial, characteristics upon mankind, inferring that man was isolated

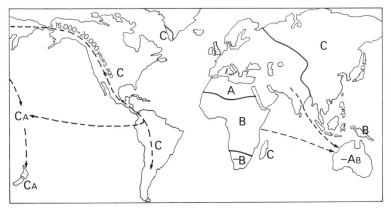

Fig. 32.—*Present distribution of the primary races and sub-races of man.*

in these environments over sufficiently long a period to ensure that subsequent changes in his habitat and conditions of living have not been able to eradicate them.

The cold northern tundras, moist grassland and steppes between the Great Ice Sheet and the desert and mountain barriers to the south (1 in Fig. 31) gave rise to the Caucasoid or "white" races (A in Fig. 32). In response to lack of sunshine, these acquired blue or grey eyes and little skin pigmentation, while, obliged to survive at very low temperatures, they developed abundant, fair, wavy hair and fewer and smaller skins pores, less open nostrils and a less everted mouth than the "black" races (*see* below).

The hot tropical forests to the south-west of the desert and mountain barriers (2 in Fig. 31) gave rise to the Negroid or "black" races (B in Fig. 32). There were races adapted to cope with excessive temperatures by such devices as an abundance of large skin pores for rapid sweating, wide nostrils and a large everted mouth, and to survive, despite high levels of solar radiation, aided by black skins and eyes, and woolly-textured black hair.

Finally, the arid grassland and steppe to the south-east of the mountain barriers (3 in Fig. 31) has been instrumental in determining the characteristics of the Mongoloid or "yellow" races (C in Fig. 32). These races developed thick dry skin, few pores, less-open nostrils and a less-everted mouth than the Negro, a broad face, straight coarse black hair, and eyes rendered almond-shaped by eye-lids in "Mongoloid folds", a device possibly developed to exclude glare and wind-borne dust from the eyes on the broad Asian plains.

Early migrations

The retreat of the Great Ice Sheet enabled *Homo sapiens* to embark upon some of the first in an extended series of migrations in which he has been engaged down to the present day. It is probable that our modern sub-races, such as the South African bushman, the Australian Aborigine or the Polynesian inhabitants of Oceania, developed out of early interbreeding between the primary races occasioned by these migrations.

It seems reasonable to suppose that the Negro peoples, indigenous to tropical regions which have an all-year-round growing season, have always been, as now, predominantly food gatherers. On the other hand, both Mongoloid and Caucasoid peoples must have always been primarily hunters, in order to survive the non-growth season prevalent throughout the non-tropical world.

With a small world population and therefore no lack of suitable land, it is probable that early man was free to roam at will, and thus did not acquire until some later date any concept of corporate or individual territorial rights. While early migrations were probably in response to climatic or other physical changes in the environment, it is likely that later migrations were prompted by "land hunger", aggravated by the pressures imposed when one migrant people made incursions upon the customary migratory routes of another.

These migrations inevitably occasioned widespread interbreeding between races, yet, so effectively had the impress of the primeval environments been stamped upon the primary races of mankind that, although there are composite or sub-racial peoples who simultaneously display the characteristics of more than one primary group, in most of them one remains dominant. This is all the more remarkable, since a truly pure race is exceedingly rare, and most peoples bear as recessive characteristics submerged traces of a very complex ancestry.

It is not only man's physiological characteristics, but also his whole language and culture—customs, beliefs, skills and technologies, as well as his patterns of social organisation—that have been spread by migration. Indeed the process has been central to the development of the world distribution of agricultural systems that exists today.

The origin and location of agricultural hearths

Human development has been punctuated by certain distinct stages: gathering and hunting; herding; possibly rather later agriculture, distinguished by the development of cultivation and domestication; and urbanisation.

Although as yet the empirical evidence provided on the subject by archaeology are inconclusive, Carl Sauer in "Agricultural origins and dispersals", *American Geographical Magazine*, 1952, argues for there having been two original centres of agriculture based upon the following prerequisites.

1. Food surpluses, possibly aided by ready availability of wild animals and fish, sufficient to permit deferment of immediate needs for experimentation in plant and animal husbandry.
2. A sufficiently wide variety of plants and animals to permit experimentation and hybridisation, a situation which in turn required a region with a sufficiently wide climatic range.
3. A suitable site at the existing level of technology. Probably this would be provided by an area of forest clearings, since early peoples could not cope with flood control in extensive river valleys, and did not have effective ploughshares required to cope with natural grasslands at this stage.
4. A race of sedentary cultivators, as the only kind of people likely to be able to guard their crops from animals.

It was through the development of agricultural technology that it became possible to support a larger population, to support more diversified economic roles, e.g. specialised crafts, and to provide the productive surpluses required for an extension of barter trade. All these developments in turn made possible the emergence of urban settlements and the more sophisticated social organisation this implies. Figure 33 illustrates the agricultural hearths, or centres of origin, and suggests the directions in which the new agricultural technologies were spread from these.

The Process of Diffusion

"Diffusion", to the geographer, is the process by which ideas, technologies, fashions, diseases or human racial characteristics—all of which may collectively be described as innovations—are spread or transferred.

A great deal of early research was undertaken by Torsten Hägerstrand at the University of Lund, in his *Spatial Diffusion as an Innovation Process*, first published in Swedish in 1953, and later in English by the University of Chicago Press in 1968. From the start his work was concerned with agricultural innovations. It has subsequently been

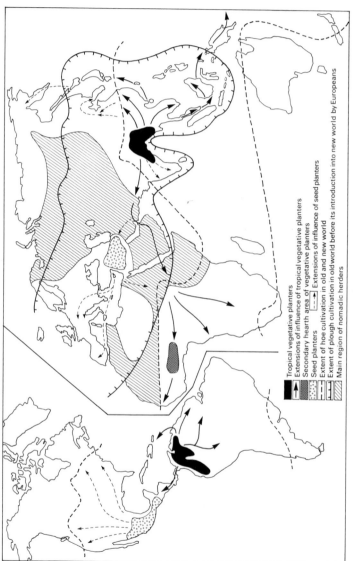

Fig. 33.— *The agricultural hearths, their lines of spread, and the extent of influence prior to European colonisation of the New World.* (This map, based upon the views of C. Sauer, E. Hahn and N. I. Vavilov, is reproduced from E. Isaac, *The Geography of Domestication*, Prentice-Hall, 1970.)

Tropical vegetative planters
Extensions of influence of tropical vegetative planters
Secondary hearth area of vegetative planters
Seed planters
Extensions of influence of seed planters
Extent of hoe cultivation in old and new world
Extent of plough cultivation in old world before its introduction into new world by Europeans
Main region of nomadic herders

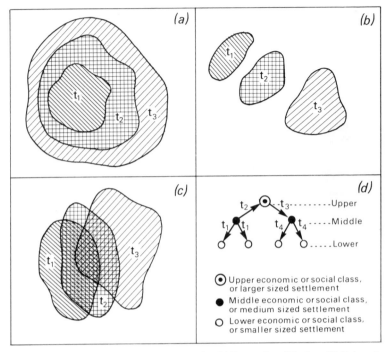

FIG. 34.—*The different forms diffusion may take.* (*a*) Expansion diffusion. (*b*) Relocation diffusion. (*c*) Forms (*a*) and (*b*) combined. (*d*) Hierarchic diffusion. Time intervals are represented by t_1, etc.

developed and applied in various contexts by writers such as Leonard Bowden and R. L. Morrill (*see* below, pp. 59 and 61).

Diffusion may take the following different forms (*see* also Fig. 34).

1. *Expansion diffusion* is the kind which occurs when an innovation spreads over an increasingly large area from its original source, as when some new discovery, which is seen at first by many as a risky speculation, is soon adopted by the majority of farmers once it has been seen to be profitable.

2. *Relocation diffusion*, where the item leaves one location for another, is the kind which occurs as a result of migration.

(1. and 2. may also occur in combination with each other.)

3. *Hierarchic diffusion* occurs when ideas are adopted first by one size of settlement, or one social or economic class of society, and spread upwards and/or downwards through other levels or classes.

Before considering the process of diffusion in greater detail certain basic elements need to be defined.

Diffusion occurs in an area. The "innovation" or item diffused is first introduced by an innovator to an adopter. The innovation follows a certain direction or path, and the units of time taken by transmission along this path between innovator and adopter (t_1, t_2, t_3, etc., in Fig. 34) are called generations. The adopter subsequently becomes the innovator and relays the innovation to others.

Diffusion, however, is not a constant process, but is subject to a number of strictures. People transmit their ideas either by public communication, which includes use of the mass media, or else privately by letter, telephone call or word of mouth. A person's field of spatial contacts is called his "Private Information Field" (P.I.F.) and research has shown that the number of contacts declines with distance (*see* Fig. 35).

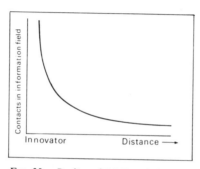

FIG. 35.—*Decline of P.I.F. with distance.*

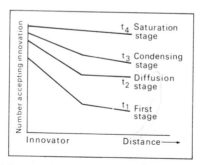

FIG. 36.—*Stages of diffusion.* First stage: most adoptions near to innovator. Diffusion stage: new innovation centres operate at distance. Condensing stage: relative increase of adoptions equal everywhere. Saturation stage: decline and cessation of diffusion.

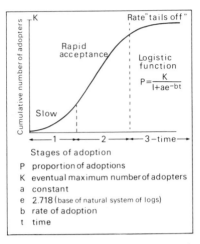

FIG. 37.—*Logistic curve marking changing rates of innovation.*

The collective field of a number of similar people is called their "Mean Information Field" (M.I.F.). Evidence suggests that this also is most effective around existing innovators, and declines with distance. This tendency is known as the "neighbourhood effect".

The effective transmission of an innovation appears to decline with distance and over time (*see* Figs. 36 and 37). Hägerstrand visualised the transmission of innovations over distance and through time in the form of *innovations-forloppet*, that is, innovation or diffusion waves. Experiments have been conducted into plotting and mapping the progress of such waves to reveal how they are affected by the slackening of acceptance rates (over time), by resistance to innovations (over distance), as well as by physical barriers which not only delay but also deflect them from the original direction of their advance.

Hägerstrand's work on the character of diffusion

Hägerstrand drew up a probablistic model to ascertain the general pattern of diffusion over time. The models subsequently described are concerned with expansion diffusion.

For his Model I Hägerstrand made the following basic assumptions which affected its practical operation.

1. There was a uniform plain over which the population was evenly spread. This was represented by a regular set of eighty-one cells with thirty people in each cell, or 2,430 people in all, and thereafter referred to as the "study area map" (*see* Fig. 38).
2. Transmission of the innovation was by simultaneous public communication to all adopters.

0	30	60	90	120	150	180	210	240
−29	−59	−89	−119	−149	−179	−209	−239	−269
270	300	330	360	390	420	450	480	510
−299	−329	−359	−389	−419	−449	−479	−509	−539
540	570	600	630	660	690	720	750	780
−569	−599	−629	−659	−689	−719	−749	−779	−809
810	840	870	900	930	960	990	1,020	1,050
−839	−869	−899	−929	−959	−989	−1,019	−1,049	−1,079
1,080	1,110	1,140	1,170	1,200	1,230	1,260	1,290	1,320
−1,109	−1,139	−1,169	−1,199	−1,229	−1,259	−1,289	−1,319	−1,349
1,350	1,380	1,410	1,440	1,470	1,500	1,530	1,560	1,590
−1,379	−1,409	−1,439	−1,469	−1,499	−1,529	−1,559	−1,589	−1,619
1,620	1,650	1,680	1,710	1,740	1,770	1,800	1,830	1,860
−1,649	−1,679	−1,709	−1,739	−1,769	−1,799	−1,829	−1,859	−1,889
1,890	1,920	1,950	1,980	2,010	2,040	2,070	2,100	2,130
−1,919	−1,949	−1,979	−2,009	−2,039	−2,069	−2,099	−2,129	−2,159
2,160	2,190	2,220	2,250	2,280	2,310	2,340	2,370	2,400
−2,189	−2,219	−2,249	−2,279	−2,309	−2,339	−2,369	−2,399	−2,429

Fɪɢ. 38.—*Hägerstrand's study area map.* Made up of 81 cells, 30 persons per cell, 2,430 in all, representing a uniform plain with an evenly distributed population. Interval 0–1 is represented by 0, thus 0–29 = 30 individuals in cell.

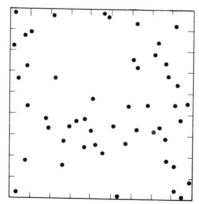

Fɪɢ. 39.—*Result of Model I, obtained from fifty adoptions.* (From T. Hägerstrand, *Innovation Diffusion as a Special Process*, translated by A. Pred, University of Chicago Press, 1967.)

3. All 2,430 adopters accepted the innovation at once, independently of each other, the acceptances being registered in random order.

Hägerstrand took a significant step by employing a "Monte Carlo" simulation (i.e. using some device to simulate chance) instead of trying to take account of "factors unknown". A simple model random order may be ascertained by taking numbers out of a hat, by rolling a dice, or, for more sophisticated models, by consulting published tables of random numbers or using numbers suggested by a computer.

To operate the model, first the cell then an individual in that cell were ascertained by use of random numbers, and this adoption was registered before proceeding to others. The resultant distribution of the first fifty adoptions is illustrated in Fig. 39. Although arrived at on a wholly random basis, there are suggestions in Fig. 39 of regional groupings which might easily (but should not) be interpreted as evidence of the operation of unknown geographical factors.

In his Model II, in order to come closer to reality, Hägerstrand modified and added to his assumptions.

1. The uniform plain of cells with an even spread of population was retained.
2. In place of "simultaneous public communication" to all adopters, transmission was by private communication between individuals in the same or different cells.
3. All 2,430 adopters still accepted the innovation immediately, although the probability of their being contacted was now restricted by distance from the innovator.
4. The first innovator was located in the centre of the study area.
5. This first innovator transmitted the item concerned in generation t_0.
6. Subsequent transmissions were at regular set intervals or generations, t_1, t_2, t_3 and so on.
7. The probability of contact between cells declined with distance in accordance with the neighbourhood effect.

To simulate the neighbourhood effect (assumption 7) for the practical model, Hägerstrand first calculated the declining probability empirically in terms of unity. In Fig. 40(a), the chance of contact with a corner cell is only 0.001, compared with 0.3 for the central cell. In Fig. 40(b), the probabilities are calculated to assess chances for each of the 25 cells out of 1,000. A Monte Carlo simulation is then employed to throw up a number between 0 and 999. In order to use Fig. 40(b) as a Mean Information Field (M.I.F.) it must be centred over the Study Area Map used for Model II (Fig. 38), so that the first innovator is located in the cell occupied by 1,200–1,229. A number then thrown up randomly, say 25, will identify the first adoptive cell as that occupied by individuals 720–749—in a cell twice removed and placed diagonally to the north-east

0.001	0.002	0.02	0.002	0.001
0.002	0.05	0.1	0.05	0.002
0.02	0.1	0.3	0.1	0.02
0.002	0.05	0.1	0.05	0.002
0.001	0.002	0.02	0.002	0.001

(a)

0	1–2	3–22	23–24	25
26–27	28–77	78–177	178–227	228–229
230–249	250–349	350–649	650–749	750–769
770–771	772–821	822–921	922–971	972–973
974	975–976	977–996	997–998	999

(b)

FIG. 40.—*Mean information field.* (a) Matrix of probabilities. (b) Sum of probabilities of all cells out of 1,000.

of the centre. If, as in Model III below, it is necessary to use the simulation to identify the adoptive individual in that cell, then a random number between 0 and 29 is thrown up. Thus, for example, 27 would indicate 746 in the cell occupied by 720–749. For the second generation, the M.I.F. is recentred over the first innovator and the first adopter in turn, so that in this generation two adoptions take place, and so the process continues.

Model II was more realistic than Model I, but Hägerstrand was bent upon achieving something closer to reality.

In his Model III the basic assumption concerning immediate adoption—third in Model I and Model II—was replaced by a built-in resistance to innovation. On the basis of empirical observations, provision was made for certain proportions of the thirty individuals in each cell on the study area map (Fig. 38) to resist innovation to a greater or lesser degree. Thus, out of thirty individuals in each cell:

2 of Class I required 1 contact for innovation to occur (low resistance);
7 of Class II required 2 contacts for innovation to occur;
12 of Class III required 3 contacts for innovation to occur;
7 of Class IV required 4 contacts for innovation to occur;
2 of Class V required 5 contacts for innovation to occur (high resistance).

This adjustment proved conducive to a more realistic model, effectively reproducing the slow ascent in stage one of the logistic (S-shaped) growth curve as displayed in Fig. 37. Thus from an initial thirty-nine generations only eight acceptances were recorded. The model also made possible greater spatial concentration of acceptances in closer accordance with reality as represented by empirical observations of the diffusion of grazing subsidies in southern Sweden between 1928 and 1944.

Effects of an irregular population distribution on diffusion

P. Haggett, in his *Geography—a Modern Synthesis*, has proposed a solution to the problem by replacing the uniform plain with its even spread of population with one displaying an irregular distribution more closely in accordance with reality. If there are different numbers of people in each cell then, in addition to probability of contact being weighted so that it decreases with distance (as in the M.I.F. in Fig. 40), it must also

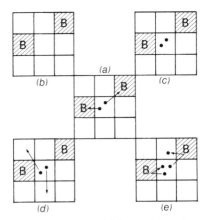

FIG. 41.—*Types of barrier to diffusion.* (a) Initial step: innovation confronted by barrier. (b) Barrier absorbs innovation and destroys the innovators. (c) Barrier absorbs innovation but leaves innovators unharmed. (d) Reflecting barrier: rejects innovation but permits another generation of it. (e) Super-reflective barrier: rejects the innovation and deflects it to the cell nearest to the innovator.

be weighted in cognisance of the differences between densely and less densely populated cells.

If we multiply the population in each cell by its original contact probability we have a joint product for each cell. The ratio between the joint product, CN, of that cell, and the sum of the joint products of all the cells in the M.I.F. ($\Sigma\,CN$), provides a weighted contact probability for that cell:

$$C_1 = \frac{CN}{\Sigma\,CN}$$

where C = contact probability, N = number of people, Σ = sum (for all cells), and C_1 = the weighted contact probability.

The need to undertake this calculation for every single cell seriously hampers the running of the model unless the services of a computer are enlisted, but it nevertheless greatly enhances its realism.

Effects of geographical barriers on diffusion

Richard Yuill tested the effects of four kinds of barrier on the diffusion of innovations through a study area map of 540 cells, using a nine-cell Mean Information Field (M.I.F.). The four types of barrier are illustrated in Fig. 41. Yuill also demonstrated how diffusion waves reform after penetrating or bypassing such barriers, despite initial diversion or delay.

Application of model of expansion diffusion to real-life situations

Leonard Bowden, in his "Diffusion of the decision to irrigate", *University of Chicago Department of Geography Research Paper 97*, published in 1965, used an adaptation of Hägerstrand's model to simulate the pattern of expansion displayed on the high plains of northeast Colorado when cattlemen there began to introduce wells in order to offset the prevalent drought conditions.

In 1948 there were only forty-one wells in the entire region, which lies between the rivers South Platte and Big Sandy Creek (*see* Fig. 42(*a*)). Bowden ran ten computer simulations through for the period up to 1962, but finding that each of these differed in detail, he averaged the results of all ten and produced a simulated map for 1962, as shown in Fig. 42(*b*). Comparison between this simulated distribution pattern showed a very close parallel with the real-life situation illustrated in Fig. 42(*c*). Bowden has since produced a projection for 1990, assuming that the model he used may still hold good for the display of future diffusion trends (*see* Fig. 42(*d*)). He has anticipated a situation in which some townships may be obliged to discontinue the sinking of wells for fear that the water reserve may be exhausted. Townships which have reached a top limit of sixteen wells are marked S for saturated on the 1990 map.

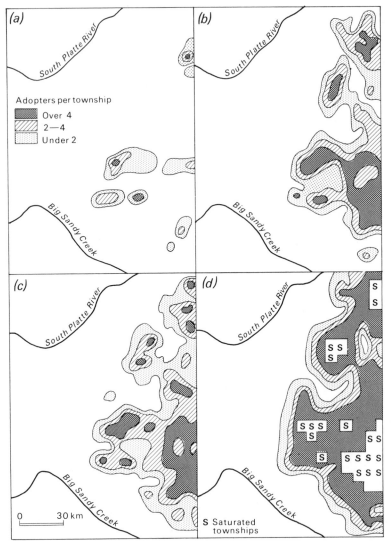

FIG. 42.—*Decisions to irrigate taken in East Colorado.* (*a*) Actual pattern of wells, 1948. (*b*) Simulated pattern of wells, 1962. (*c*) Actual pattern of wells, 1962. (*d*) Projected pattern of wells, 1990. (From L. W. Bowden, University of Chicago Department of Geography Research Paper No. 97, 1965.)

Other examples of expansion diffusion that might be considered include such situations as those in which knowledge of vegetative reproduction must have been carried by migrants from the south Asian agricultural hearth shown in Fig. 33; in which the agricultural co-operative system has been adopted in various parts of the world; or in which strains of maize and rice developed by Western technology have been adopted in Monsoon Asia during the so-called "Green Revolution".

A model and real-life example of relocation diffusion

R. L. Morrill, in his "The Negro ghetto—problems and alternatives", *Geographical Review*, Vol. 55, 1965, made a model to explain the process by which the black ghetto is extended. He envisaged this as a process of spatial diffusion, in which the black man is the active agent, the white man an agent of resistance, and a house purchase is analogous to an innovation. Morrill's Simulation Model was close to that of Hägerstrand in that it was concerned with the diffusion pattern and not with individual movements.

On the basis of empirical observation, Morrill made several significant assumptions:

1. that the Negro ghetto in Seattle was in every way, economically and culturally, inferior to the adjacent communities it threatened;
2. that the threatened expansion was encouraged by rising population of the ghetto (estimated by him as about 5 per cent every two years), expansion of the neighbouring Central Business District, and Central City Transport developments;
3. that black migrants tended only to move a few blocks from their previous home.

This last observation prompted Morrill to devise a M.I.F. equivalent, which he called a Migration Probability Field, and which could be superimposed over the migrant's block.

Migrant movements were to be determined by drawing random numbers, and by observing the following simple rules.

1. If the random number indicated a block with Negro occupants already, then movement was to be completed immediately.
2. If there were no Negro occupants, no move was to be made but the contact was registered.
3. If a subsequent contact was made in the same or subsequent two-year period in the same block referred to in 2. above, then the move was to be completed forthwith.

Morrill observed that near the ghetto the number of white buyers fell within a distance of only five to seven blocks by 96 per cent. He noted that fear-mongering estate agents encouraged panic selling amongst white owners so as to deflate house prices.

Morrill used his model to simulate the growth of Seattle's ghetto region for the periods 1940–50 and 1950–60. His simulation for 1950–60 was remarkably close to reality (*see* Fig. 43).

Relocation diffusion is applicable in agricultural contexts, e.g. in

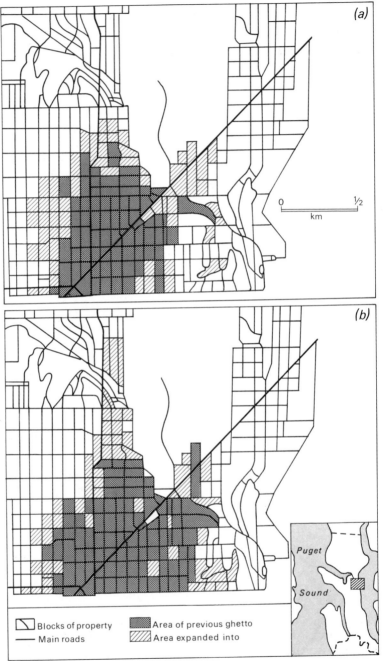

FIG. 43.—*Expansion of Seattle's ghetto, 1950–60.* (a) Simulated expansion. (b) Actual expansion. Inset' Position of study area. (Reproduced from R. L. Morrill, "The Negro ghetto: problems and alternatives", *Geographical Review* Vol. 55, 1965, with the permission of the American Geographical Society.)

Texas during the 1880s when the extensive livestock system, represented by the cattle ranchers, was ousted out of all but the most arid regions by the extensive cereal cultivators, in much the same way as that in which the white inhabitants described by Morrill were displaced.

The process of hierarchic diffusion

This process was defined on p. 52. It remains to explain that construction of a model of hierarchic diffusion is different in that, in place of a contact field in which contact probability declines with geographical distance, a new exponential contact field has to be devised to take account of socio-economic distances between different sizes of settlement or classes in society.

This is relevant to some situations in agriculture, although the process has little bearing upon distribution of the agricultural systems. An example which might be cited is how, during the eighteenth and early nineteenth centuries, various agricultural innovations adopted by aristocratic landowners spread down to their tenants, through the incentives offered by the land agents who managed the estates, or through the publicity afforded by such exponents of the new methods as Tull, Young and Cobbett. This downward form of hierarchic diffusion is also referred to as cascade diffusion.

Evolutionary Change

Although the present-day distribution of agricultural systems owes much to the process of diffusion, and to the physical barriers which have so often restricted it, evolution of agricultural systems has also taken place in many regions with minimum, if any, external influence. Thus one or all of trade expansion, industrial or technological advance, or population growth can stimulate such changes in the prevalent system of agriculture that there is evolutionary development into another more advanced system. Alternatively, a region may be faced with some environmental challenge to which it must either find a technological answer or die. Thus whereas primitive man failed to find a solution to his overgrazing of the drying Sahara savannahs of prehistory, modern man with the incentive provided by the discovery of oil may manage to reclaim the desert, and in so doing create a new agricultural system by means of his technology. Such a process may be illustrated by considering a model of change.

A model of evolutionary change

William Kirk, in his "Problems of Geography" (*Journal of the Geographical Association*, No. 221, Vol. XLVIII, Part IV, 1963, pp. 364–6), has identified two very significant concepts which must be taken into account in developing a model of change, namely those of the phenomenal and behavioural environments.

The phenomenal environment (P.E.) is made up of not only natural phenomena but also environments altered and in some cases almost entirely created by man. The behavioural environment (B.E.) may be defined as a "psycho-physical field in which phenomenal facts are arranged into patterns or structures (*Gestalten*) and acquire values in cultural contexts". Taking these concepts as a starting-point, the following model of change suggests itself (*see* Fig. 44).

1. *The phenomenal environment* is made up of the whole physical environment and the cultural overlay which has been imposed upon it by man's occupancy over thousands of years, and which in some cases has almost entirely replaced it. It may therefore be seen as the summation of geographical conditions to date, upon which a whole superstructure of socio-economic, cultural and political influences, making up a related system known as the behavioural environment, is sustained.

2. *The behavioural environment* represents the real world as it is perceived by its inhabitants, who have acquired their perceptual patterns (*Gestalten*) and a system of cultural values from it as a part of their nurture. In other words, man acquires his physical character from the phenomenal environment, but it is the behavioural environment that invests him with the perceptual equipment that colours his interpretation of his environment. Indeed, he sees not the phenomenal environment but the behavioural environment he has been conditioned to see. It is this that colours his view of his problems, suggests the solutions, and limits both his ability to carry them out and his perception of their effectiveness.

3. *Human attitudes.* The great mass of people tends to behave and to think collectively. People are aware of inadequacies in their environment which lead to dissatisfactions, which are coloured by the values and beliefs they have acquired from their environment. Their protestations and complaints give rise to pressures for change.

4. *Pressures for change.* These calls for help are felt in the phenomenal environment, where they stimulate the emergence of decision-makers.

5. *Decision-makers* are the "natural leaders" in society—individuals with initiative and drive, and ability to think and act independently of the "mass". Nevertheless, they too have acquired their nature and nurture from the same P.E. and B.E., unless, as sometimes happens, they are "introductions" from a different P.E. and B.E., when their impact is all the more profound because it involves innovations from their different environment.

6. *The decision-making process* involves the development of solutions to problems which may or may not be implemented as intended, but eventually events of change occur.

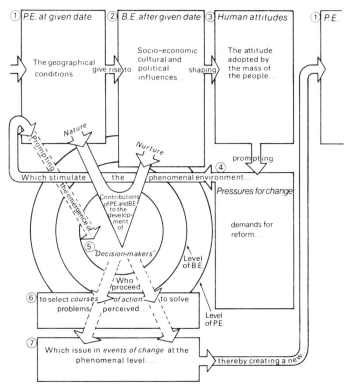

① P.E. at given date | ② B.E. after given date | ③ Human attitudes | ① P.E.

The geographical conditions... give rise to | Socio-economic cultural and political influences... shaping | The attitude adopted by the mass of the people...

Nature | *Nurture* | prompting

Which stimulate the phenomenal environment... | Pressures for change

④

Contributions of P.E. and B.E. to the development of | demands for reform...

⑤ "Decision-makers" | Level of B.E.

Who proceed

⑥ to select *courses* of *action* to solve problems / perceived... | Level of P.E.

⑦ Which issue in *events of change* at the phenomenal level... | thereby creating a new

FIG. 44.—*A model to demonstrate the processes underlying evolutionary change.*

7. *Events of change* modify the phenomenal environment, although they were chosen in relation to the behavioural environment as perceived. The solutions and their consequences can only be appraised in similar relative terms, which is why the solutions and their consequences upon the phenomenal environment are rarely as effective as intended.

8. *The new and modified phenomenal environment* sustains a correspondingly new B.E., with its own appreciable shortcomings, which now serve to set in motion a new sequence of change.

In Fig. 45 this model is used to illustrate the evolutionary changes which brought about the development of nomadic herding and subsequently extensive rotational livestock farming among the erstwhile primitive nomadic hunters of the north-east Siberian tundra.

Human Migrations of the Past 500 years

The development of the agricultural hearths (*see* Fig. 33) may have contributed to the emergence of three main Old World concentrations of

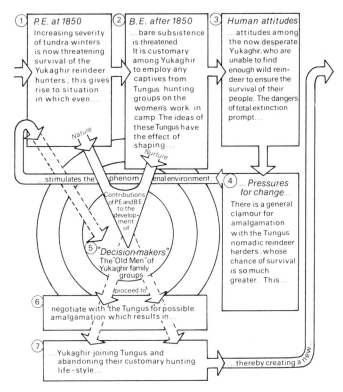

FIG. 45.—*Application of model of evolutionary change shown in Fig. 44.* Used to illustrate
the evolution of the inhabitants of north-east Siberia from a hunting economy,

population in about the time of Christ. These were a European Centre,
which originally emanated from the Middle East, with 45 million
people, a North China Centre based upon the Hwang Ho Valley with
75 million people, and an Indian Centre based upon the Indus Valley
with 12 million people, out of a total world population of 300 million.

After another 400 years the Roman Empire, which until then dominated the European Centre, collapsed, but following a period of severe
economic and political regression during the Dark Ages, the late
medieval period was marked by commercial expansion, and the population of Western Europe in particular displayed a recovery. Since 1500
Europe has displayed a most remarkable economic resurgence.

During the past 500 years, human migrations, which have mainly
emanated from Europe or from areas occupied by people of European
origin, have helped to determine the present distribution of the world's
agricultural systems (*see* Fig. 46). These migrations, voluntary and
forced, involved some 95 million people, of which it is estimated that

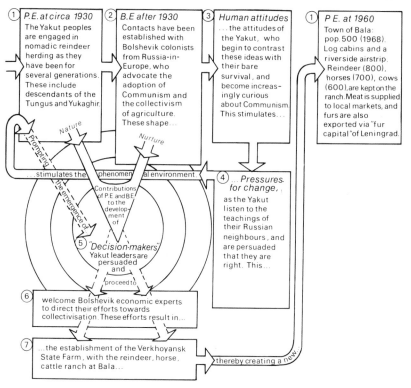

① P.E. at circa 1930
The Yakut peoples are engaged in nomadic reindeer herding as they have been for several generations. These include descendants of the Tungus and Yukaghir.

② B.E. after 1930
Contacts have been established with Bolshevik colonists from Russia-in-Europe, who advocate the adoption of Communism and the collectivism of agriculture. These shape...

③ Human attitudes
...the attitudes of the Yakut, who begin to contrast these ideas with their bare survival, and become increasingly curious about Communism. This stimulates...

① P.E. at 1960
Town of Bala: pop. 500 (1968). Log cabins and a riverside airstrip. Reindeer (800), horses (700), cows (600), are kept on the ranch. Meat is supplied to local markets, and furs are also exported via "fur capital" of Leningrad.

Nature
Nurture
Prompting
the emergence of
...stimulates the phenomenal environment...

④ ... Pressures for change,
as the Yakut listen to the teachings of their Russian neighbours, and are persuaded that they are right. This...

Contributions of P.E. and B.E. to the development of

⑤ "Decision-makers"
Yakut leaders are persuaded and proceed to

⑥ welcome Bolshevik economic experts to direct their efforts towards collectivisation. These efforts result in...

⑦ ...the establishment of the Verkhoyansk State Farm, with the reindeer, horse, cattle ranch at Bala...

thereby creating a new

through nomadic herding, to sedentary extensive livestock farming in little more than a century. *See* Chapter VI for details of these changes.

nearly 70 per cent were Europeans, 20 per cent Africans and the remaining 10 per cent Asians. As a result of these migrations, Old World crops such as rice, wheat and coffee have been introduced into the New World, and New World crops such as tomatoes, tobacco and potatoes have found their way into the Old World.

Below are summarised the main phases of migration, and their implications for agriculture.

1450–1820: from the age of discovery to the end of the age of mercantilism

Europeans, first the Portuguese and Spanish and later the French, Dutch and English, opened up the newly discovered lands of the Americas, Africa and South Asia. They did this by establishing trading stations along the coasts of China, India, West and East Africa, and Malaya to trade for products such as spices, tea and silk. They also established plantations in Brazil, Mexico and the Caribbean and later

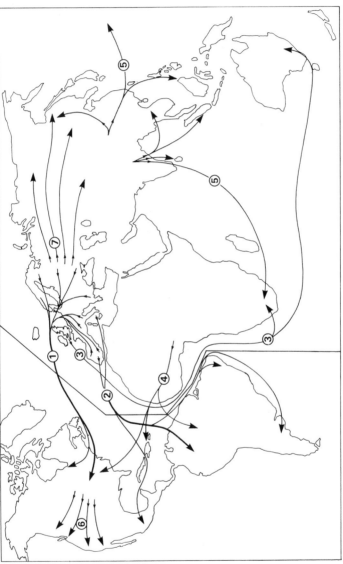

Fig. 46.—*Main intercontinental migrations since the sixteenth century.* 1. Europeans to North America. 2. South Europeans to Central and South America. 3. British to Africa and Australasia. 4. African slaves to North America and West Indies. 5. Indians and Chinese, mainly within Asia. 6. Americans westwards in North America. 7. Western Russians into Siberia. (From W. S. and E. S. Woytinsky, *World Population and Production: Trends and Outlook.* Copyright © 1953 by the Twentieth-Century Fund, Inc. Figure 27, page 68. Reproduced with permission.)

in East Africa for the production of tropical products such as sugar and spices, and later coffee and cacao. As these activities expanded so the supply of indigenous labour proved inadequate, and slave labour was imported from West Africa.

In the mid-latitudes European colonists established agricultural settlements. In North America these were predominantly Dutch, French and English, and later in Australia and New Zealand they were mainly English. A study of the farm settlements of North America will display local diversities of settlement pattern and farm practice which stem from the wide range of European minorities that were involved.

1820–1918: the age of industrialisation and imperialism

In this era of rising European population and industrial expansion the demand for tropical produce such as coffee, tea, jute, cotton and rubber, and mid-latitude produce such as meat and wheat, encouraged an extension of agricultural activities into the vast continental interiors, aided by technological advances such as the establishment of railway networks, faster transport made possible by steamships, and the introduction of refrigeration which made possible the importation of perishables such as meat and fruit.

A growing number of Europeans emigrated to help in the development of the great plains and prairies of North America, the Argentinian pampas, the South African veldt, the Australian outback and the Canterbury Plains of New Zealand. In tropical regions the growing demand for plantation products led to increased importations first of African slaves and later of indentured Indians. European countries began to vie with one another for trading concessions in the coastal regions of China.

Since 1918

The Russian Communists in Europe have colonised Siberia, and the Japanese and Chinese have engaged in policies for commercial and territorial expansion in the South-east Asian region.

Summary: Combined Influence of Physical and Human Factors

It will be useful here to look back at Fig. 20 to relate all that has been said to the world distribution of the main agricultural systems.

Primitive nomadic gatherers and hunters, cultivators and herders continue to survive in the less hospitable regions of the earth, having been displaced from most more productive areas. Such primitive peoples, because of their limited ability to adapt the environment to their needs, provide the most effective demonstrations of geographical determinism. Contrasts in their geographical environments account for pronounced regional contrasts between these peoples and between their agricultural practices.

Figure 20 also demonstrates that, whereas cultivators and gatherers are usually found in tropical forests and savannah regions where there is an all-year-round growing season and sufficient precipitation to sustain a subsistence supply of food, hunters and herders, who depend on animals which can themselves survive under a wider range of geographical conditions, occupy regions of steppe or hot desert, or the taiga-tundra zone which extends across the high latitudes of Laurentia and Eurasia.

Examination of Figs. 31 and 32 reminds us that man's early migrations, which urged forward the Mongoloid peoples from their formative area on the steppes of South-east Asia to take possession of the Americas, but also gave rise to such sub-races as the Australian Aborigines (composite Negro and proto-Caucasoid), Hottentot (proto-Negroid) and Polynesian (Mongoloid with a tincture of Caucasoid blood), must in addition have carried primitive agricultural techniques from the Old World to the New World. Such migrations therefore brought about relocation diffusion, but in addition there is evidence of a slow spread of these self-same techniques outwards from the agricultural hearths by a process of expansion location, barred sometimes by such geographical barriers to diffusion as great ranges of mountains such as the Himalayas, and great deserts such as the Sahara.

In the past 400 years the economic changes associated with the rise of capitalism, discovery, colonialism and mercantilism, and since the eighteenth century the Industrial and Agrarian Revolutions, have all contributed to the present distribution of the major agricultural systems.

In this period Eurasia has continued to be the source of innovations which have subsequently been diffused throughout the world. The main factors which have shaped the distribution of her own agricultural systems have therefore been evolution and expansion diffusion. Thus the primitive sedentary rotational system with fallow periods, popularly called "open field", which evolved in the continental conditions of the eastern part of the North European Plain but reached Jutland and England by expansion diffusion, has subsequently developed towards the intensive rotational mixed system and the co-operative systems. Likewise, true Mediterranean-type agriculture is the evolved response to the climatic conditions of the littorals of the Mediterranean Sea. A similar evolutionary process lies behind the development of the agricultural practices of the people of north-east Siberia (*see* Fig. 45).

The contrasted modern development of China and the Indian subcontinent are also worthy of consideration. In both, the peasant intensive subsistence system of agriculture had evolved in response to the monsoon climate, which had also determined regions where rice was the predominant crop or otherwise. Throughout history, however, the Himalayas had formed a barrier to the diffusion of cultural influences of any kind between them. Whereas China had experienced Marxist influences, therefore, the Indian subcontinent, having been under British influence or rule for the previous 250 years, was not disposed to follow the same path, when, in the late 1940s, the Chinese set up a Communist form of government.

China's collective system of agriculture was certainly the result of expansion diffusion of Marxist innovations in those early days, although throughout the last twenty years or so evolutionary tendencies and the decline of Russian influences have led it to diverge along distinctive Chinese lines. The slower evolution of agriculture in the Indian subcontinent, although efforts have been directed towards the application of scientific ideas from Western technology, does not display so sharp a break with what went before.

In contrast, the other continents display the effects of relocation diffusion brought about where immigrants of European descent have imposed their alien cultures and technologies on the indigenous peoples, or, as in the New World and Australia, have almost entirely displaced or destroyed the original inhabitants along with their cultures over large areas.

Thus the North American continent is peopled almost exclusively by people of North European origins, who despite differences in detail, as, for example, between French and English in Canada, introduced and imposed North European agricultural practices. Likewise, there has been a predominant Spanish influence in South and Central America, while South Africa has experienced Anglo-Dutch and Australia and New Zealand a British influence. As a result the original peoples, Red Indians, South American Indians, Hottentots, Aborigines and Maoris respectively, have been assimilated, or segregated to become cultural anachronisms in their own lands, which now reflect the cultural imprints of their conquerors instead.

The different geographical conditions encountered by these European immigrants in their adopted lands have led, in general, to two distinctive responses. The European colonists of mid-latitude regions have achieved greater independence from their countries of origin. Obliged to evolve new systems of agriculture better suited to the vast continental extents they now occupy, they have developed commercial extensive systems of agriculture to provide surpluses for export to industrial Europe. These systems include commercial extensive livestock farming, notably cattle ranching and sheep farming, and commercial cereal cultivation, mainly engaged in production of wheat. The demarcations of these two systems have been climatically determined, since cereal farmers tend to oust the ranchers into the drier regions where the more profitable cereals are at a disadvantage. Modern movements towards mixed farming practices in these regions have also been the result of evolution.

The European colonists of tropical regions, instead of practising European styles of agriculture themselves, have sought to impose their methods on tropical peoples. In some cases they organised the forced migration of tropical peoples from one region to another. The plantation system was a complete innovation, unprecedented in Europe, yet incorporating modes of labour organisation and agricultural skills and practices developed in Europe and adapted to the new situation.

The forced migrations of tropical peoples—Negro slaves from West Africa to the U.S.A. and the West Indies, and indentured Indians to East Africa—were involved in a process of relocation diffusion. The forcible imposition of plantation agriculture on peoples who were accustomed to primitive tillage and herding was also an example of relocation diffusion, since indigenous practices were entirely disrupted.

The present day is seeing the diffusion of two antagonistic cultures, capitalism and communism. Both are evolving in response to changing social and economic conditions, with the communists advocating collectivisation, and the capitalists moving, sometimes reluctantly, towards the adoption of co-operative methods or at least towards the development of agricultural associations.

Finally, modern commercial horticulture and fruit growing have grown up in response to a rapid growth in market demand for such products, made possible by improved transport and marketing techniques, but still restricted to regions with suitable soils and climates. Similarly, factory farming is an evolutionary response to a rapid growth of commercial markets, but has adopted scientific technology as its means of providing plentiful cheap food in regions where there is insufficient agricultural land available. In other words, these are two modern systems which have evolved in response to rather different economic pressures.

It is clear from the very generalised summary above that the physical and human factors have been extremely complex and inextricably mixed. Therefore any attempt to assess their relative contributions to the agricultural pattern at this world scale is quite abortive. In considering specific regions in Part Two, however, an attempt can and will be made to assess these influences as they have operated at the local level.

Chapter V

The Economics of Agriculture

THIS chapter is concerned only with the economic aspects of agriculture as a form of production, particularly where these are of geographical significance. For wider consideration of economic theory the reader is recommended to consult any reputable introductory text on the subject, possibly one of those listed in the bibliography.

The Factors of Production

The economist distinguishes three main factors of production:

1. *land*, which in the agricultural context means literally land, but without development of any kind;
2. *labour*; and
3. *capital*, a broad category including the sum of all the man-made resources acquired by deferment of past consumption for the purpose of applying them to future production.

Some distinguish also *entrepreneurship*, which includes qualities such as leadership, managerial skill and business acumen.

Each of these factors is necessary, although if one is in limited supply, more of one or both of the others may be substituted. Thus in the extreme case of a cattle ranch, *a large area of land* is employed for the production of beef and hides, but with only a very small application of labour, and with capital confined mainly to livestock and relatively small stocks of equipment. In contrast the alluvial river valleys of Monsoon Asia are traditionally farmed in very small land units, little or no capital is required for the minimum of seeds, livestock or implements employed, but there is a *heavy application of cheap, plentiful labour*. Both differ fundamentally from the situation found under factory farming. This system has developed in built-up regions where practically no land or sufficiently cheap labour is available for agriculture, yet there is a large and profitable market for products such as meat, dairy produce, fruit, flowers and vegetables close at hand. In this situation, *by applying heavy inputs of capital*, in order to provide the most sophisticated scientific technology, and employ automation and the principles of industrial management, agricultural production more than competitive with conventional farming systems is achieved, despite only negligible applications of labour or land.

Diminishing returns and economies of scale

The size of an agricultural unit is often determined by socio-economic factors at least partially outside the control of agriculture, whereas a crucial factor in determining the proportions in which labour and capital are applied to production on a given area of land is the level of technology available. If, despite the technology being available, insufficient labour and/or capital is applied to an area of land, then this constitutes a waste of its potential productivity. Increased applications of labour and/or capital will initially step up output more than proportionately with the additional marginal inputs (increased returns). In the subsequent stages the rate of increasing output will diminish, first until

TABLE 2

Non-proportional returns to a variable factor of production

No. of men	*Total product* (*tonnes*)	*Average product* (*tonnes*)	*Marginal product* (*tonnes*)
1	8	8	8
2	24	12	16
3	54	18	30 (*a*)
4	82	20.5	28
5	95	19	13
6	100	16.7	5 (*b*)
7	100	14.3	0 (*c*)
8	96	12	−4 (*d*)

NOTE: Points (*a*), (*b*), (*c*) and (*d*) are shown on the graph in Fig. 47.

it is constant with the marginal inputs, then until it is relatively less with each additional marginal input, until finally there is a zero increment for each additional unit of input (diminishing returns). In Table 2, it is assumed that one variable factor, labour, is applied to a fixed area of land, with a constant amount of capital. The effects on returns are represented graphically in Fig. 47.

Temporarily such decreases may be offset by changes in technology, making possible the application of more capital or labour to land, but there is a limit to such increased returns, and diminishing returns soon set in again. Any substantial increase in output can only be achieved by increasing the entire scale of production—more land, labour and capital—which even at the same level of technology will make economies of scale possible, up to a point where the inevitable diminishing returns set in again. A hypothetical upper limit of scale where maximum output is achieved at minimum cost may be regarded as the optimum size for a productive unit in that particular type of agricultural system and at that existing level of technology. Table 3 illustrates these points.

TABLE 3

Returns to scale

No. of men	Area of land	Total product	Percentage growth of enterprise	Percentage increase in total product
4	10	54	—	—
8	20	120	100	122.2
12	30	198	50	65.0 (*a*)
16	40	264	33.3	33.3 (*b*)
20	50	315	25	19.3 (*c*)
24	60	360	20	14.3

NOTE: The scale of production is increased at the point on Fig. 47 where average product per unit of labour is greatest. Increased returns persist until (*a*); constant returns occur at (*b*); and once again diminishing returns set in at (*c*). (Based on C. F. Stanlake, *Introductory Economics*.)

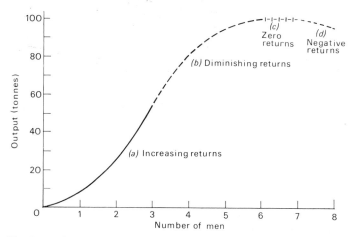

FIG. 47.—*Diminishing returns*. Graphical representation of Table 2. (Based on C. F. Stanlake, *Introductory Economics*.)

Economies of scale at different levels

Although we tend to visualise agriculture in terms of applying the other variable factors of production to a fixed allotment of land, it is of course of equal importance that the best productive use should be made of any one factor, by combining it with the most appropriate proportions of the others. Thus, a relatively expensive skilled workman, such as a good cowman, or a costly piece of capital equipment, such as a combine harvester or a grain drier, may be wasted if underemployed. There is some evidence, for instance, that many British farms, often those in remoter regions, are over-capitalised. In order to utilise fully any factor of production it must employed at an appropriate scale of operations.

To avoid the situation of being over-capitalised, and yet continue to enjoy the benefits of employing skilled technicians, expensive scientific techniques and scientific appliances, farmers in many countries engage in various degrees of association. Many Danish farmers, for example, are committed to the use of agricultural co-operative societies. British and Norwegian farmers, while certainly more geographically isolated from one another than their Danish counterparts, are perhaps less willing to surrender their independence, and have experimented, showing less commitment, with agricultural associations and machine rings. Such associations make it possible to organise agricultural production at two levels of optimum size, and enjoy economies of scale at both. There is a large optimum size for many specialised services or for purely administrative functions, and a smaller optimum size appropriate to practical day-to-day operations of crop and livestock husbandry on the farm. In any case the small sizes of farm may be the outcome of socio-economic or physical restraints outside the control of the farmers concerned.

Land as a Factor of Production

Ever since prehistory the population of the world has been increasing, slowly until modern times, but with an increasing acceleration in the last few hundred years (*see* Table 4).

TABLE 4

Growth of world population

Date	Population estimate	Probable rate of increase per century
6000 B.C.	less than 5 million	6–10 per cent
A.D. 1	250 million	2.5–5 per cent
A.D. 1650	470–545 million	65 per cent
A.D. 1980	4,000 million	300 per cent
A.D. 2000	6,000–7,000 million	

NOTE: Based upon Emrys Jones, *Human Geography*.

There can be no comparable increase in the land area available for agricultural production, various reclamation techniques notwithstanding, and so an upper limit of world agricultural land must be reached in due course. There is a possibility that instead of fishing, which is comparable to hunting, fish farming could become general, and that food crops could be produced by cultivation of the continental shelf and even of the surface of the oceans. Nevertheless, such developments can only delay man's inevitable arrival at the ultimate limit of food production. Thus, although so far the problem has been offset by technological innovations, Malthusian threats of "diminishing returns", when no amount of additional inputs will issue in increased food supplies from the world's limited productive resources, must eventually forebode starvation for countless millions of people.

As yet this threat at world level is not imminent, but already some effects of land hunger and diminishing returns are making themselves felt in countries containing areas of high population density, particularly where such vast concentrations of people contain a large proportion directly dependent upon agriculture for their livelihood.

In India in 1891 the average farm-holding was 1.0 ha in extent, in 1941, 0.6 ha, and in 1976, 0.4 ha. If it is borne in mind that large areas of India are sparsely populated, the above statistics of average farm size are the more remarkable, and when seen in relation to India's rapidly growing population (in the single decade after 1961, the population grew by some 25 per cent) it is clear how the situation is deteriorating. The tiny farms are fragmented; often a dozen farmers must tend a field of less than a quarter of a hectare. Under conditions such as these, economies of scale are not practical, and the increased yields achieved have been thanks solely to technological innovation. Even then, the growing population threatens to absorb entirely the hard-won increases of about 2 per cent per annum.

In industrialised countries, where the bulk of the population is not dependent directly upon agriculture, the problem of land takes another form. Agricultural technology and a tendency towards larger farm units have made possible unprecedented yields which have foiled the threat of diminishing returns. Food has remained comparatively cheap—surprisingly so when the loss of land to uses other than agriculture is taken into account.

This loss of agricultural land is a function not simply and directly of rising population, but of higher living standards which have led to urban industrial sprawl, the burgeoning of motorways and rail networks, the construction of reservoirs to meet growing needs for domestic and industrial water, the setting up of reafforestation schemes, tourist sites, public utilities such as airfields, scheduled historical and archaeological sites, and government requisitioned areas for research establishments, military training grounds, etc., all of which have reduced the land available to agriculture. In other words, the population growth in such areas has been comparatively small, but the need to satisfy the insatiable demands of the affluent societies has been inordinate.

Ricardo's theory of economic rent

David Ricardo published his *Principles of Political Economy* ("political economy" was the old term for economics) in 1817. In it he expounded his theory of economic rent as applied to land, on the grounds that land, unlike other factors of production, was in fixed supply. This in fact is not necessarily so in any particular region or for some specific use. Neither did Ricardo attempt to apply the principle to other factors of production, i.e. labour, capital or entrepreneurship, to which it may be equally applicable given inelasticity of supply, e.g. the skilled craft of a brain surgeon or an artist.

Economic rent may be defined as the payment made for any factor of production in excess of the sum required to keep that factor in the mode of production, or prevent it transferring to another. Applying this definition to a non-productive use of land, e.g. residential, a person may seek to pay enough to live in a high-cost region to avoid higher costs of travelling backwards and forwards to his employment, or because he enjoys living in the property, which has a beautiful view, and he wants to discourage his landlord from selling it to make way for a golf-course.

Economic rent is therefore an important consideration in determining land-use. For example, it is relevant to situations in which there are different qualities of land, since until changes in technology make production under different conditions possible, there may be a very inelastic supply of some particular type of land in any particular location. It is also relevant to the distributional patterns adopted by competing land-uses, based upon their position and corresponding production and marketing costs in relation to market prices. This applies at the scale of a single agricultural productive unit, where land-uses are determined in part by situation in relation to the farm nucleus, or at the scale of the agricultural region, where land-uses are related to regional market centres.

These applications of the theory of economic rent were in fact arrived at independently of Ricardo by von Thünen (see Chapter I). The principles shown graphically in Figs. 48, 49 and 50 are therefore also demonstrated in the regional situations already cited to illustrate von Thünen in Figs. 2–10.

In addition, economic rent is an important economic factor in determining land values and land-use where agriculture may come into direct competition with other, non-agricultural land-uses. An example is when the very survival of farms on the periphery of large urban centres is threatened by land development schemes, and in order to retain their land farmers resort to all kinds of intensive systems within what has been called the "market-gardening and dairying zone".

The function of economic rent in this context, and indeed within the entire urban environment, is illustrated in Fig. 51. The steep rent gradient between successive land-uses, whether urban or agricultural, reflects the successive increments of economic rent for increasingly desirable locations nearer to central employment and market oppor-

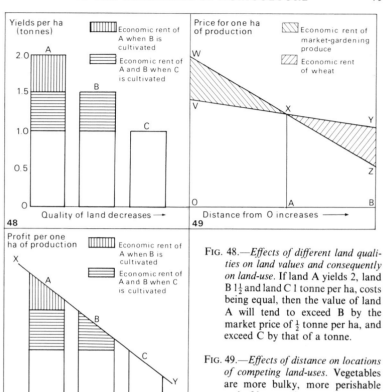

FIG. 48.—*Effects of different land qualities on land values and consequently on land-use.* If land A yields 2, land B 1½ and land C 1 tonne per ha, costs being equal, then the value of land A will tend to exceed B by the market price of ½ tonne per ha, and exceed C by that of a tonne.

FIG. 49.—*Effects of distance on locations of competing land-uses.* Vegetables are more bulky, more perishable and of lower value per unit weight than wheat. Thus, VWX, the economic rent of market-gardening produce over wheat, rapidly declines with distance, while XYZ, wheat's economic rent, favours its cultivation further away from market O beyond point A.

FIG. 50.—*Effects of distance on land values and consequently on land-use.* Profit derived from the sale at market O of 1 ha of wheat grown at A will be OX, but the profits decline at B and C (sloping line) because of the increasing transport cost to O. As a result, the economic rents (shaded portions) accrue to locations A and B if land at C is cultivated. (Based on M. Chisholm, *Rural Settlement and Land Use.*)

tunities. It is therefore easier for more profitable or intensive land-uses, such as building development in the "broker belt", or market gardening, to expand outwards than vice versa, since it is improbable that a less-profitable system could afford to compete with a more profitable system such as market gardening.

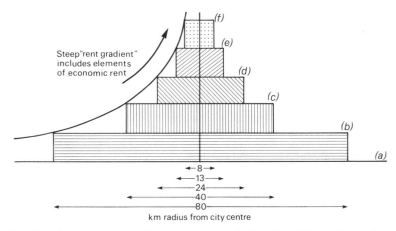

Fig. 51.—*Competing land-uses in and around a hypothetical city.* (*a*) Types of agriculture appropriate to outlying region. (*b*) Market-gardening, factory-farming and dairying belt. (*c*) Suburban, high-class, low-density residential. (*c*) Inner, high-density residential. ((*c*) and (*d*) also have some modern "light" industrial sites.) (*e*) Mixed commercial, industrial, residential functions in "down-town" area. (*f*) City-centre area with prestige properties.

Government intervention and land reform

Governments sometimes intervene in the process of adjustment by the forces of supply and demand, in order to undertake land reforms for economic and socio-political reasons. They are often concerned to increase land-holdings to nearer the optimum size where technological changes have made this larger than previously, and yet social customs or economic restraints have delayed the natural adjustment which should have followed. Thus the *Remembrement* programme in France is designed to assemble larger, compacted holdings in place of tiny, fragmented 12 ha holdings which were the legacy of the famous Napoleonic Code Civil, which at its inception in 1804 had been designed to establish holdings of desirable size. The government in Italy has engaged in similar policies.

In the Netherlands we have a unique situation of a country that has "created" by reclamation a considerable area of new land for agriculture. The post-war holdings have been planned smaller than those of pre-war years, partly because demand from farmers exceeds supplies available, and partly because improved planning and technology has made such smaller holdings viable.

In many countries, democratically motivated governments favour the existence of a large number of small, independent farmer proprietors, as is reflected in policies in such diverse situations as Denmark and Malaysia. In contrast to this, extreme socialist and communist regimes are often bent upon land reforms for political and ideological reasons, based upon egalitarianism. In their view, land should be confiscated by the state, and the estate-owner and small-farmer/proprietor classes are both deemed undesirable. This is not to be confused with the reallocation of estate lands to the peasantry that may take place following the establishment of a military dictatorship, where it may be interpreted as a means of destroying the power of a landed ruling class and gaining popular support at a single stroke.

Labour as a Factor of Production

Where socio-economic and political pressures bring about changes in population structure, there may be changes not only in the proportions in each age group, and in the relative proportions of the sexes in each age group, but also in the occupational structure of the entire country. This in turn may have repercussions on the availability of agricultural labour, and upon the scale and character of the markets for agricultural products.

Urban industrialisation and rural decay

Developing countries, such as those of Monsoon Asia and the African or South American continents, display a pattern of change which is familiar to students of eighteenth-century British economic history. Peasants are constrained to leave the expensive and unproductive land of their villages of birth in the expectation of more money and food in the new expanding industrial cities. However, they only serve to swell the queues of unemployed seeking casual labour and living in the appalling shanty towns which have sprung up in squalor and malnutrition on the fringes of the established cities, such as Calcutta, Johannesburg and Lima. For the inhabitants, crime and vice are the ugly consequences of detribalism, and of the breakdown of the social values of the settled agrarian societies which the migrants have left behind. The rural areas that have been abandoned as a result of this exodus to the towns are consequently no longer able to sustain traditional methods of agriculture. The remaining, often ageing, population has to find means of revitalisation for the rural economy.

All is not lost, however, because with fewer agricultural workers to manage, but also to depend on, the land, more land is available to support a man and his family. There is the prospect of larger consolidated holdings together with new urban markets where surplus produce can be sold, so there is promise of commercialised modern farming to take the place of primitive subsistence. Nevertheless, in the meantime, the costs in social suffering, with insufficient capital and few technological skills to make the transition quick and painless, are extremely high.

Although it might be supposed that the United Kingdom was well past the consequences of urban industrialisation, there is ample evidence to the contrary. In the nineteenth century, even after the parliamentary enclosure movement had brought about the rural exodus to the towns, there were still numerous farm labourers and servants, not to mention many more engaged in seasonal work such as ploughing, reaping and winnowing, thatching or shearing. However, the twentieth century has seen an accelerated decline in their numbers. Thus in 1900, 9 per cent of the British labour force was engaged in agricultural employment, but in 1976 only 2 per cent. In the post-war period agricultural employment has fallen from 837,000 in 1950 to 434,000 in 1973.

Evidence of rural decay in Wales includes an ageing and declining population with attendant decline in the potency of village society with its chapels and choirs; agricultural land taken over for reafforestation for lack of farmers, with consolidation of the surviving farms into larger, more economic holdings; and, most serious of all in the eyes of her people, a slow erosion of the Welsh language and culture as the dwindling Welsh population is encroached upon by English-speaking residents. Similar developments among the Gaelic-speaking communities of the north-west Highlands and Islands of Scotland are aggravated by a tendency for children from remoter settlements who have received their secondary education in an urban centre away from home to choose not to return.

In English villages particularly there has been a growing tendency for the young people to seek more congenial employment in nearby towns, in preference to work on local farms. Farmers have, in response, tried to make agricultural work lighter and hours shorter, and to pay wages nearer to those offered in the town, but they have usually found it advantageous to apply more capital, in the form of labour-saving machinery, to agricultural production, leaving much of the former agricultural labour force of the village to commute daily to the town by car. Village decay has been further aggravated by the invasion of urban commuters seeking pleasant rural homes, and turning the once living, caring village community into an empty shell—"a collective attempt at private living" as Mumford has put it. Examples of such "rurban" settlement are those Chiltern villages which have become merged into London's so-called "stockbroker belt".

Labour and different agricultural systems

Availability of labour is a crucial factor in determining the system of agriculture practised in a region. Thus in the period when the Europeans colonised the vast mid-latitude and tropical grassland regions of North and South America, South Africa, Australia and New Zealand, the indigenous populations were found to be, not merely sparse, but made up for the most part of primitive gatherers and hunters. Furthermore the settlers themselves were not in sufficient strength to do more than practise extensive subsistence agriculture in a pioneer situation. Gradually, with the development of improved communications, by rail and by sea, with clippers and steamships, a little later with the introduction of refrigeration, the agriculture of such regions acquired rich industrial markets in Europe, and was soon to adopt commercial extensive systems of operation. In the twentieth century urbanisation and industrialisation have increased the domestic population of such regions, leading to a degree of intensification in agriculture, and the development of mixed agriculture in place of monoculture. Throughout their development, and down to the present day, the great grassland regions have always suffered from lack of labour. They have therefore led the world in applying a high level of capital to production, pioneering large-scale mechanised farming in order to maximise production with a minimum of costly labour.

This has been in strong contrast to the traditional agriculture of Monsoon Asia, where labour is plentiful and cheap, and capital and land in extremely short supply. Until economic development is able to provide alternative sources of employment away from the fields, India has to take care to avoid certain technological changes which serve only to create technological unemployment, and which threaten to accelerate rather than stem the rural decay which arises from rural depression and underemployment. The answer appears to lie, here as in other underdeveloped countries, in the development of "bridging technologies", i.e. innovations which lighten work but do not radically reduce labour requirements. A more long-term solution lies in the development of hydroelectric schemes, whereby the reservoirs and the operation of electric pumps can together help extend the arable area by irrigation, and also provide electric power by which village life may be enriched, making possible domestic heat and light, education by radio and/or television, and the establishment of rural crafts which diversify rural employment, improve living standards, and, more important, provide the means of removing excess labour from the land. The emergence of a new consumer market in the regenerated villages as well as the industrial towns can eventually bring commercial agriculture into being, and new larger farm units as well as industry would help to generate the capital required for the all-round improvement of the agricultural economy of rural India.

It is clear from the above examples that the availability or otherwise of labour has been the key to their past agricultural development, and is central to this and their entire economic future prospects.

Capital as a Factor of Production

The application of more capital and less labour to land has been a persistent theme in English agriculture ever since the end of the Middle Ages, when Tudor landowners began to apply more sheep and fewer labourers to their holdings, and so set in motion the early phase of private enclosure. The Industrial and Agrarian Revolutions from the eighteenth century onwards provided additional motivation in the expanding urban industrial markets for agricultural produce, and in the technological improvements in agriculture and transport which, given the appropriate capital investment on the part of the landowner, held out for him the possibility of larger agricultural surpluses which could profitably be sold in the new markets. Thus followed the great era of parliamentary enclosure, which was to affect the Midlands region in particular. Enclosure transformed the appearance of the English landscape by investing it with a chequerboard of smaller enclosed fields. The effective arable unit in the open-field system had been tiny contiguous strips into which the open fields were divided, which inhibited innovation and experimentation. Following enclosure the farmers were better able to introduce new methods, such as segregation of livestock for selective breeding, or the use of horse-drawn machinery, on their consolidated and well-defined holdings. Outside the Midlands, in the north and west or the extreme south-east of England, land had already been enclosed for some considerable time.

Effects of increased capital input on field size

In Britain the twentieth century, particularly the second half of it, has seen a further drift of labour from agriculture, aided no doubt by enhanced mobility of labour (*see* p. 82). The application, in its place, of greater proportions of capital has brought about changes in agricultural practice and with them changes in the aesthetic appearance of the countryside, not least among these being the introduction of new mammoth field machines and a corresponding movement to enlarge the fields themselves.

Examination of old estate maps will indicate that the average size of fields, in areas where enclosure took place early, has been increasing steadily throughout the last two centuries at least. An example of this can be found in T. Bolton and P. A. Newbury, *Geography through Fieldwork: Book 3* (Blandford Press, 1970), which contains maps of the Brookgate Estate in Kent in 1718 and in the present day.

Recently, however, the process has been accelerated, particularly in the cereal-growing counties of eastern England, but even in parts of Devon and Cornwall, the very stronghold of the tiny Celtic fields. Initially the intention was to enlarge the fields and increase their accessibility to the great new combine harvesters. It was soon realised, however, that the hedges and the great trees which had surmounted them had in fact occupied or overshadowed quite substantial areas of potentially productive farm-land. Fewer hedges actually meant larger farms—and larger output without any increase in the total holding.

The new landscape looks bare and strangely unfamiliar, but already more practical misgivings have been expressed. These relate to soil erosion, and to exposure of crops and livestock to high winds, where so much shelter has been removed.

Natural and Economic Risks which confront Agriculture

Without the benefits derived from trade and "cash" markets, medieval English farmers, like their counterparts throughout the "Third World" today, were subsistence farmers—growing crops and producing livestock to provide for the needs of their own small local community. Small surpluses, when they occurred, might go to the relief of a less fortunate village, or might be exported to serve a remote urban market such as London, which ceased to be self-sufficient at an early date.

In England the open-field system, which was introduced from the Continent by the Anglo-Saxons, evolved in response to the vagaries of the unpredictable cyclonic climate, becoming "mixed", i.e. dependent upon a range of plant and animal products instead of concentrating heavily upon cereal cultivation, so as to offset climatic hazards. The risks the farmers of the "Third World" have to combat are even greater, however, which probably accounts in no small measure for the underdeveloped state of these predominantly agricultural countries. Thus the Monsoon climate brings excessive rains and flooding, or droughts which, equally disastrously, result in crop failure. Other hazards include plagues of locusts, earthquakes, severe tidal waves which inundate coastal regions, and severe epidemics which ravage entire populations. It is little wonder that, confronted by such frequent and various disasters, the peasant of Monsoon Asia, work as he may, can never escape from his state of chronic indebtedness.

The adoption of modern "cash-crop" systems by such people is heralded by them, and by observers throughout the world, as ensuring for them a new prosperity by engaging in trade. What is often minimised is the fact that it also exposes them to a new range of economic hazards when they become dependent upon world markets, for indeed they are often overdependent upon world markets in a narrow range of agricultural products for their survival. Examples of the hardships attendant upon such overdependence include the impact of disease in bananas upon the Jamaican economy, or of depressed rubber prices on the Malay economy as in the 1930s.

One of agriculture's chief handicaps in relation to industrial markets is its inelasticity, i.e. its inability to respond quickly to fluctuations in market demands. Demand on the part of industry for agricultural products may be very elastic, i.e. subject to sudden adjustments to take account of changing circumstances such as substitution by margarine or soap manufacturers of one vegetable oil source for another, or substitution of margarine for butter, and vice versa. Similar substitutions occur from time to time, when prices vary in relation to one another, between, for example, rubber and synthetics, natural and synthetic textiles, one type of meat and another, or meat and soya substitutes.

The production prices of agricultural commodities depend upon such things as world oil or cattle fodder prices, and so the farmer cannot always produce at competitive prices. Similarly the consumer's purchasing power is susceptible to external factors, e.g. wage and employment levels. At the same time agricultural supply, certainly in the short term, is exceedingly inelastic, that is, a farmer is committed for many years ahead when he invests in permanent plant such as silos, or in pedigree herds of cattle. He is also certainly committed for a year ahead when he prepares to grow a crop or to rear a certain number of animals. Any unforeseen change in his production costs or in market prices can therefore be disastrous to the farmer.

If farmers are not to farm extractively, to the long-term detriment of their land-holdings and basic capital resources such as farmhouse or barns, leading to soil exhaustion, neglect and obsolescence, then they need to enjoy the security necessary to make reinvestment worth while. In other words, they require protection from the worst consequences of impoverishment brought about by natural or economic disasters lying outside their control. Various solutions have been found to meet the farmers' needs. Under capitalism, where private enterprise is a cherished ideal, farmers often engage in self-help, through agricultural co-operatives, as in Denmark, or looser types of association. Capitalist governments have also found it necessary for many years to provide farmers with support in the interests of national self-sufficiency, or for the sake of political stability. The main devices adopted by governments have been:

1. improvement grants, along with sponsoring of research, dissemination of information to farmers, etc.;
2. programmes designed to reduce the real costs of production while increasing demand;
3. crop-limitation programmes designed to reduce supply and so raise farm prices;
4. subsidies, which raise the producer's income while lessening the cost to the consumer;
5. guaranteed prices, which involve the government in either buying up stocks of the commodity not required by the public at the guaranteed price, and retaining them until some subsequent period of shortage, or buying the entire output of the producers and absorbing any losses incurred at the current resale price.

The specific devices adopted by the E.E.C. are considered in Chapter XVII.

Technological and Social Changes affecting Agriculture

In addition to the natural and economic risks considered above, against which governments have often found it necessary to afford protection to the farmers, there are the more predictable changes that occur, such as improvements in transport which help create or expand existing markets and so weaken or destroy others, or changes in public tastes and consumer priorities.

Mention has already been made of the ways in which the clipper, the steamship, the steam locomotive and refrigeration between them opened up vast continents to commercial agriculture, but also disrupted the markets of farmers in several European countries. In recent years increased airfreight space has led to international air trade in livestock, dairy produce and market-garden produce. In addition, the development of "freightliner" trains, and "on–off" freight-loading terminals between the U.K. and the European mainland, have helped to change the patterns of trade, and consequently of agricultural production, in the E.E.C. countries.

It is subsequently observed (*see* p. 270) that changes in the priorities of the affluent society, particularly as regards the place given to food, have encouraged the farmer to seek in technology the solution to the problem of producing ever more, and relatively cheaper, food on a diminishing area of agricultural land. In addition, socio-economic factors have brought about changes in tastes and consumer habits to which the farmer has been obliged to respond, for which the following few examples must suffice.

Today's smaller family requires a much smaller size of meat joint. In addition, the more effective heating systems of today, not only in the home but also in the workplace, and the extension of the sedentary range of occupations, have led to a decline in the necessary average calorific intake. Finally, there is now a strong association in the public mind of heart disease and high blood pressure with obesity and a high intake of animal fat. All these factors have affected modern views on diet, and with these changes in consumption have come corresponding changes of production. The great joints of mutton and beef, the wether, the free-range hen, have all been sacrificed to the demand for young, tender cuts of meat with very little fat. Likewise the dumplings and heavy puddings have been at least partially supplanted by the slimming products of horticulture. Market gardening, factory farming and modern dairying have all grown up and expanded in response to these changes in public taste.

Part Two

THE AGRICULTURAL SYSTEMS OF THE WORLD

Chapter VI

Primitive Nomadic Hunters and Herders

Introduction

IN the next two chapters primitive systems of agriculture are considered, and it is appropriate here to indicate the order of presentation.

As already stated in Chapter IV, early man subsisted by gathering and hunting before true agriculture emerged. Rarely did any people subsist by the practice of only one of these activities. Gathering was the predominant occupation of tropical peoples who could rely on an all-year-round growing season, although such people hunted a little as and when the opportunity occurred. Similarly hunters all over the world practised gathering as a subsidiary activity during their short "season of plenty". Such gathering and hunting, in the early period universal, is today confined to primitives who, although they still occupy large areas of the earth, are comparatively few in number and have been ousted from all but the most inhospitable of regions.

Primitive nomadic hunters have in many cases evolved into herders. Having acquired the art of domesticating animals, they are able to retain them near at hand and so reduce the insecurity of the chase.

Similarly, as shown in Chapter VII, many primitive gatherers have learned to propagate and improve plant species, and their success as cultivators has been reflected in higher living standards and correspondingly higher populations, which have obliged them to evolve from nomadic to sedentary forms of cultivation. In order to achieve and maintain the consistent level of soil fertility required for permanent or semi-permanent cultivation to be carried out, they have developed rotation, fallow and dressing with organic fertilisers.

The world distribution of primitive nomadic hunters and herders

An examination of Fig. 20 will indicate that present-day hunters are largely confined to the tundra regions of Laurentia and Eurasia, along with such smaller pockets as those occurring in the Kalahari and North Australian Deserts. Herders occur in a great belt of hot desert and steppe regions extending right across North Africa, across Saudi Arabia and right across South Central Asia. All these regions experience severe extremities of temperature and suffer from limited precipitation.

Used in the present context, "primitive" implies the crudest of hand-fashioned implements, and efforts directed to obtaining the bare necessities of life. "Nomadic" existence is necessary for primitive hunters because the herds of herbivores upon whom they depend are themselves migratory, driven in higher latitudes by changing seasonal temperatures to seek better pasturage, and in tropical regions by seasonal shortages of water. Herders, while enjoying the greater security that comes from constant access to their livestock, must likewise pursue a migratory search for grazing for their animals. They have achieved their higher living standards because of their possession of domesticated animals.

The domestication of animals

It appears that the dog (now believed to be derived exclusively from the wolf) was first domesticated by hunters at the close of the Ice Age. However, true domestication by herders came much later, and was, the evidence suggests, a very sudden technological development with several species of animal being domesticated simultaneously. Animals may be useful to man in one or more of the following main ways: to provide meat, to produce milk or for use in draught. In order to be successfully domesticated a species needs some inborn affinity with man, and must respond to comfort, require a minimum of care and be readily bred in captivity. Most such animals are of Old World origins, and include the pig, poultry, cattle, sheep, goats, the horse, the ass and, less extensively, the yak, the camel and and reindeer. Domestic animals of agricultural significance of solely New World origin are confined to the alpaca and the llama. Forde, in his *Habitat, Economy and Society*, contends that Old World animal domestication was begun in the Near East between 5000 and 4000 B.C. and that within two to three thousand years all the animals amenable to domestication were already being kept successfully.

Examples

Having considered the world distribution and evolutionary relationship between hunting and herding, two regional examples are discussed in the rest of this chapter. These are the bushmen of the Kalahari, possibly the world's most primitive surviving hunting peoples, and the reindeer hunters of the north-east Siberian tundra, most of whom have gone over to herding and yet more recently extensive livestock farming.

The Bushmen of the Kalahari

The Kalahari environment

The South African plateau, at a height of 1,000–2,000 m, experiences a subtropical climate, and most of it a summer rainfall maximum. The eastern half receives heavy rainfall which gives rise to dense forests nearer the coast, but the forests soon degenerate into parkland, then bush savannah, thorn scrub, and finally sand or stony desert. This latter is the Kalahari Desert, extending from the Okavango Swamps and Lake Ngami in the north to the Orange River in the south (*see* Fig. 52).

The rainfall is subject to wide variations, permanent water being confined to a few river-bed depressions or hollows where the permanent water-table reaches the surface. In wetter years the "sands are covered with grass... luxuriant bush, clumps of trees... Even the spaces between the satin grass are filled with succulent melons and fragrant cucumbers and in the earth itself bulbs, tubers, wild carrots, potatoes, turnips and sweet potatoes grow great with moisture and abundantly multiply." (Laurens Van Der Post, *Lost World of the Kalahari*.) The area is rich in game, including antelope, giraffe, ostrich, zebra, elephant, all varieties of cat including lion and leopard, hyena, jackal, and tiny creatures such as lizards, frogs, bees and locusts which supplement the bushman's diet. The problem for primitive people is therefore a lack not of food, but of readily accessible water.

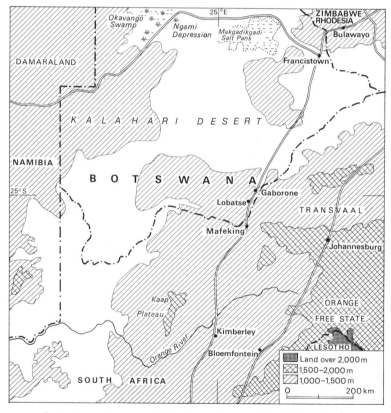

FIG. 52.—*Environment of the bushmen, hunters of the Kalahari Desert.*

In the long, dry, winter season the bushmen have to search for sufficient moist water-holes and sufficient of the now much-depleted vegetal cover in order to survive. The fact that the wild game are also desperate to locate water and tend therefore to congregate around the dependable water-holes is the bushman's chief security, since, while plant food is hard to come by, animal foodstuffs are now comparatively plentiful, and so the bushman is predominantly a hunter.

The bushman

The bushman is, according to Forde, Negroid but without "projecting mouth, everted lips", etc. Laurens Van Der Post remarks that with his broad face, apricot-brown skin and slant eyes the bushman was often designated by his ancestors as the "Chinese-person".

Racially the bushman displays such distinctive features as a tendency to "steatopygia", i.e. excessive fatty tissue on the buttocks particularly among the women (a device of nature for the conservation of surplus fats and carbohydrates), and the semi-erection of the male penis from infancy on, a feature from which the bushman derives his name for himself—"Qhwai-xkhwe".

Historical and archaeological evidence suggests that the bushman once extended over a far larger area than he does today, and his distinctive "click language" is recognisably similar to that of some hunters in Tanzania and other parts of the East African Plateau, pointing to a common origin with these peoples. However, 500 years ago full-statured Bantu-speaking Negroes were already expelling the bushman from better-watered pastures in what is now Zimbabwe Rhodesia, Transvaal and Eastern Botswana, while the Hottentots, who were cattle herders and more advanced, although racially akin, were making incursions into his territories from the south. The Dutch settlers who penetrated and expropriated his lands after 1652 looked upon him as vermin and set upon his extermination by organised "shoots", so that he was very soon expelled or eliminated in the south. Consequently only a relatively small area of his former territories in northern Namibia and Western Botswana remains to him, and even here he is harried by the taller Negro with whom he has sometimes learned to co-exist in a kind of symbiotic servitude. His response is usually one of peaceful avoidance and withdrawal.

It is possible that there are only a few thousand true bushmen (as opposed to the "tame" bushmen found in association with white or Negro) surviving and retaining the primeval way of life. Forde contends that even when they retained moister territories their civilisation was never more elaborate. Possibly the only concession to their deteriorated circumstances has been a reduced population density. The vast area each group requires relates to its capacity to sustain life during the winter drought season. A family band some 20–100 strong is the basic social unit, and a number of such bands with a similar dialect will live adjacent to one another. The larger dialect group will rarely meet except at water-holes during the drought season. It has a loose territorial significance but cannot be regarded as a political unit. While individuals may make visits to kin, trade or hunt in adjacent territories, and while groups may occasionally collaborate for a joint hunt, there are no more permanent associations.

Within the family band the senior male, "Old Man", will select the site for an encampment, usually some way from a water-hole for fear of alarming the animals. He will indicate his choice by kindling a fire, and the women will proceed to construct their dome-shaped grass huts on their stick frame. The women also go out to gather, bringing home roots, berries and tubers, and insects and such small game as iguanas, lizards, frogs and tortoises, tucked in the folds of their shoulder cloaks. They also collect the firewood and water, transported in an ostrich eggshell or a bag made from a dried buck's stomach.

The men go out daily to hunt, alone or with a son who is learning the craft. Usually they return the same night, unless caught up in a protracted pursuit. The hunter will carry a bow and arrow, and will select a salt-lick or water-hole calculated to attract game. In approaching his selected quarry he may mimic the cries of young of that species. All the time he is approaching closer to the windward side, until he is close enough to release his arrows, dressed with poison made from plant juices, the bodies of dried grubs or spiders, or the poison sacs of venomous snakes. Such poisons are effective but slow, and in the subsequent long pursuit magic may be employed to slow down the wounded beast. There is always the danger that a hyena or vulture may snatch away the animal if contact is lost. In the wet season, an animal may be pursued into a mud-hole, or if in excessive drought a beast sheds its hooves, it is more easily run to earth and clubbed to death. Individual hunters attach baited nooses to bowed saplings, or fence off a water-hole, substituting one which they have dug out and poisoned by sinking weighted euphorbia branches in it.

A larger hunting group may construct a brushwood fence for miles across a valley, or dig a great pitfall several yards in extent along the path to a water-hole, covering it with brushwood and driving larger game into it. Each man hunts for his immediate family, but in a joint hunt the proceeds are fairly shared. A hunter depending on others to bring in his kill retains the valuable hide and sinews, and directs the sharing out of the meat. While hunting a man may earmark things he has found for future use, marking the bees' nest, patch of roots or ostrich nest with his arrow, thereby asserting his ownership. Caches of food and stores of water are sometimes left to meet future contingencies. In extremities another may avail himself of these, but convention requires that he notifies the owner or else a feud may ensue.

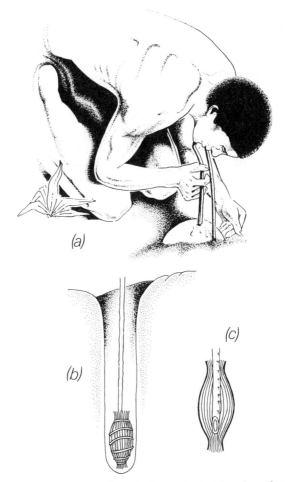

FIG. 53.—*Bushman syphons water from the deep under-desert sands at the sip-wells.* (*a*) Water is drawn up the hollow reed and allowed to trickle from the corner of the bushman's mouth into an ostrich eggshell. (*b*) Section of sip-well showing grass filter against drift sand. (*c*) Section of filter. ((*a*) from photograph by Laurens Van Der Post, *Lost World of the Kalahari.* (*b*) and (*c*) from C. Daryll Forde, *Habitat, Economy and Society.*)

Bushman technology, although limited by materials at hand, is highly ingenious. Wood is used to construct implements such as drills to kindle fire, digging and throwing sticks, and spears. Tree bark is twisted into cord. Reeds gathered from around water-holes provide arrows or the sucking tubes used at the sip-wells (*see* Fig. 53). Plentiful game provides bone for bows or, rubbed down, for fashioning arrowheads; hides for clothing or carrying bags; or sinew for bowstrings. In the absence of traded iron implements, flint can be fashioned into fur scrapers, drills or arrow points. Whole ostrich eggshells serve as water carriers or, when fragmented are polished by the women, to make beads. Their cylindrical necklaces, the so-called "bushman beads", have an origin lost in antiquity, and serve as jewellery, or a medium of exchange for spearheads, millet, tobacco, hides and knives, when these are purchased from the Bantu tribesmen to the north. Other products acceptable in trade include honey and wax, feathers, skins and ivory.

If we assess a civilisation by its culture and not simply by its technology, it must be said that the bushman has a rich fund of oral literature and instrumental music, and a repertoire of religious dance, as well as once being possessed of a rich tradition of painting lively hunting scenes on rock faces, very reminiscent of those at Altamira in northern Spain. (The latter tradition is now, sad to say, defunct.) This stresses the essential humanity even of this most backward pre-agricultural people.

The Reindeer People of the North-east Siberian Tundra

The environment

The Yukaghir once inhabited the plains north of a mountain arc formed by the Verkhoyansk Range to the south-west, and the Khrebet Cherskogo and Okhotsko Kolymskoye Ranges on the south and south-east margins respectively (*see* Fig. 54). The plains are drained northwards by the rivers Yana, Indigirka and Kolyma with their tributaries , and with their northerly situation aggravated by an isolation imposed by

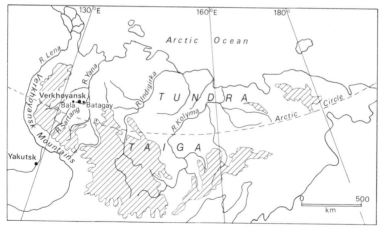

FIG. 54.—*Territories occupied by reindeer-hunting and herding peoples of north-east Siberia.*

mountains 700–1,000 m high, they experience the most severe temperatures in the world. Mean January temperatures at the mouth of the Yana are around −22.5°C. This is, however, nothing like the world's lowest temperature, recorded just south of Verkhoyansk as −67.5°C. There are only 70–80 days per year frost free (June, July and August), and the total rainfall is about 135 mm per annum, practically all in the brief summer season. The almost continuous sunshine of this brief summer makes possible a short-lived but remarkable profusion of plant growth, though handicapped by failure of the earth to thaw below a depth of 1 m (the permafrost), restricting plant growth both by physiological drought and by the fact that roots cannot penetrate the ice layer. The limited winter precipitation falls as hard snow, which is blown into deep drifts against hills and trees by the Arctic gales. The southern hills are clad in stone-pine, larch, birch and alder, which also follow northwards the more sheltered valleys, while further north the tundra supports only scanty dwarf birch and arctic willow, in addition to the small berry-bearing bushes, mosses, sedges and lichens which are the support of wildlife during the brief tundra summer.

Animals roaming the tundra in summer include the reindeer, which search out mosses, lichen, fungi and shoots. Some reindeer remain all year round on the tundra, while others seek out forest clearings in the more southerly taiga, until the irritating mosquitoes again herald approaching spring, and drive them northwards from the forest borders. Other creatures on the tundra in summer include the polar hare, elk and wolverine, and such birds as ptarmigan, duck and geese. In the southern taiga dwell the black bear, the mountain sheep, musk-deer and the squirrel. Both in early spring and later in the summer, the rivers are teaming with fish.

The Yukaghir

The Yukaghir reindeer hunters who once ranged this vast region now survive only as a remnant along the upper Kolyma basin. They are Mongoloid, short stocky people with flat yellow-brown narrow faces and straight coarse black hair. Their traditional way of life, indeed their very survival, depends upon the reindeer. In summer the Yukaghir travel by raft and canoe upon the numerous rivers and lakes. Their summer tents are of reindeer hides spread upon light poles, like the conical tepees of the North American Indians. The women gather the fruit and berries of the tundra, while their menfolk either fish with net or line, sun-drying the spring surplus in preparation for the ensuing winter, or snare the abundant marsh fowl. In autumn the men set about packing the tents on to birchwood dog sledges for the long trek southwards to winter quarters, near the forests along the slopes of the southern hills. Now sinew-strung bows and arrows or rifles obtained by trade are employed for hunting reindeer, musk-deer, elk and mountain sheep when the hungry animals are too wary to reach at close quarters. Sometimes the reindeer are stampeded down converging avenues of stones or posts to be ambushed by concealed hunters at the far end. In spring, reindeer, poised splay-footed upon the surface of the tundra marshes to graze, can be alarmed and bogged down in their haste to escape. For winter tents the fur is retained on the hides for warmth, and the footings flanked with snow to exclude draughts. Reindeer hides are also used in the making of tunics, aprons, leggings and long boots for both sexes, while the Yukaghir also depend upon the animal for its meat, its sinew for lines and bowstrings, and its bone for implements such as fish gorges. In short, so great is the people's dependence upon the reindeer that its seasonal migrations dominate their way of life. Even their religious thinking is permeated with the need to placate the spirit protectors of the animal.

The Yukaghir hunting group, rarely more than 100 strong, is directed by the eldest able-bodied man in the group, who allocates tents, boats, nets and sledges, and organises seasonal migrations as well as hunting, fishing and gathering sorties. Only clothing and weapons rank as personal property. The produce of hunting and fishing are "pooled", and allocated by the wife of the old man according to need.

There has been a long-standing enmity between the Yukaghir hunters and their reindeer-herding neighbours, the Tungus to the south and the Koryak to the east. Tungus herders have invaded Yukaghir territories and Yukaghir war-parties have attacked Tungus domestic herds. Where prisoners have been taken, the Yukaghir have humiliated them by setting them to do "women's work" around the camp. In Fig. 45 it was hypothesised that it may have been through contact with such captives that, in times of extremity, the Yukaghir have felt it necessary to join up with the Tungus herding peoples.

The Tungus

The northern Tungus are related to a more extensive group of peoples whose territories extend southwards as far as Manchuria and Outer Mongolia, and eastwards to the Sea of Okhotsk. The economic activities of the Tungus vary within this vast region; those with whom the Yukaghir came into contact combine reindeer herding with hunting and fishing. Physically the northern Tungus are like the Yukaghir, but taller.

During the summer they take their reindeer northwards on to the tundra to take advantage of the better grazing afforded by the summer vegetation there. They also induce their reindeer to eat a little fish and meat. The pasture in any one place is rapidly exhausted by their large herds, and since it may take up to a few years for such pasture to recover from excessive grazing, the group must move on every few hours. The huge area required for the grazing of a single animal, possibly about 10 km², leads to severe competition between groups for the limited grazing available.

Tungus domestic reindeer are larger than their wild cousins. Most of the herd (ranging from a few dozen to several hundred per family) consists of does and "followers", a few bucks for breeding, and a number of geldings (desexed bucks) retained for meat and draft purposes. The does can yield about half a litre surplus of thick sweet milk a day in addition to suckling their fawns. They are only capable of carrying half the load (74 kg) of a gelding or buck. The reindeer can be ridden (over the shoulders), loaded up in a pack train, or employed to draw a sledge.

The Tungus are dependent upon their herds for much of their food, clothing and shelter, but very few groups even attempt to live entirely off their herds, which are susceptible to disease epidemics and the depredations of wolves. Most northern Tungus therefore supplement the produce of their own herds by hunting and fishing, and it has been in pursuit of these activities that they have fallen foul of the Yukaghir. Hunting methods employed include the use of tame reindeer as decoys, or employing a tame reindeer among the wild herd concealed behind a screen on a tiny toboggan.

The Tungus trade in fur-pelts with Russian and Russianised sedentary Tungus for guns, iron implements, tea and tobacco. They also barter Maral deer hartshorn with the Chinese, who value it as a medicine, for scrap and bar iron, which their smiths use to beat out rough implements. Between themselves, however, there is no trade or barter, although reciprocal lending in time of need is universal.

Within a tribe possibly up to a thousand or more strong, there will be two or more clans, each of a semi-autonomous nature, with names such as "Woodpecker's Noise" or "Poplarwood Cradle". Each clan may include several hundred family units of four to ten people, coming together briefly for joint exploitation of the summer's good things, but operating independently throughout much of the year, the men hunting and trapping, the women tending the herds and even striking camp by themselves, moving to a pre-arranged rendezvous to meet their husbands again. Men may share specific tasks in the care of the herd such as sawing off the antlers or gelding, leaving the day-to-day care to their womenfolk, who rarely hunt or fish. Older men share herd management with the women. In short, the customary division of labour is flexible in practice.

Since the 1920s Russian Communist colonisation of Siberia has affected life in the region (see Fig. 45). The partial assimilation of the Yukaghir by their more technically skilled Tungus neighbours described above has represented for the Yukaghir an evolution towards herding, in contrast to which Russian colonisation of Siberia has been achieved with an element of coercion and economic persuasion of the indigenous peoples.

It is difficult to assess to what degree the different indigenous peoples have undergone amalgamation. Either the Tungus have been displaced from their foothold in the basin of the Yana north of the mountains by other similar people who like them originated further south in the vicinity of Lake Baikal, or there has been a further phase of assimilation between Tungus and Yakut. The Yakut were also Mongoloid ethnically, but Turkic in tongue. They brought with them cattle and horses from the south, and in efforts to keep these on tundra pasture have forced them to eat meat and fish. The Yakut talk of their reindeer herds as "foreign cattle", they brand instead of earmark their reindeer, and as their horses have dwindled, so they have taken to saddling and riding their cattle. It is this people who have now adopted extensive livestock farming in one-time Yukaghir country in the basin of the River Yana.

The Bala reindeer ranch

Bala is part of the Verkhoyansk Kolkhoz (State Farm) situated south of Verkhoyansk on the River Sartang, a tributary of the River Yana. The small village of Bala houses 500 Yakut, and is situated in the broad wooded valley with lush pasture on the valley floor, and fir-covered hill slopes beyond. This pasture has been developed by the careful levelling and draining of the once marshy valley floor. One-storey log-cabins afford the only kind of construction possible on such shallow foundations. A concrete foundation would crack because of uneven stresses at the level of the permafrost. Arctic winds are excluded by packing the cracks between logs with moss and plastering the outside with mud. Roofs, with so low a rainfall, are flat. The rooms inside can be heated with a centrally situated "Russian stove", fuelled with logs, by the simple expedient of constructing all walls to fall short of the ceiling and having no interior doors.

The village lies upstream from Verkhoyansk, but the river is rarely navigated. There is a "winter road" (i.e. sledge path) to Batagay, 160 km away. Moscow is 6,500 km away, and the nearest part of the Trans-Siberian Railway is 1,600 km to the south. The problem of such natural isolation has been largely overcome by provision of an airstrip on the valley floor some 300 m from the village. With two aircraft a day and subsidised agricultural wages the farmers are well able to afford the favourable airfares, pegged at one-third of international rates.

The ranch at Bala is about the size of Wales. There are no sown crops, although quick-maturing varities of wheat such as those introduced in the Canadian tundra would be practical. Likewise, although some vegetables are cultivated, the official policy is to discourage such production as too labour intensive. Some vegetables are imported, although the Yakut have always been able to maintain robust health upon an all-meat/milk diet. Some haymaking for winter fodder is carried on along the valley floor.

The ranch keeps approximately 800 reindeer to provide meat, hides and furs, but milk is not considered economic to produce. Reindeer also provide for draught purposes around the ranch, taking winter fodder to livestock, or bringing in timber or hay.

There are also 700 horses of a very small breed, light brown or white in colour, not large enough for riding, and chiefly kept for meat. This is tender and bland in flavour, and regarded locally as a delicacy. Horses are grazed in groups of twenty under mounted escort along the valley floor to conserve the pasture.

A herd of 600 cows is used to provide, in addition to calves for beef, an excellent rich milk which, in the absence of nearby markets is used to make *smetana*, a slightly sour condensed milk which is easier to transport and considered a great delicacy amongst the Russians. In the severe winters the cattle have to be stall-fed in large byres.

There is a small silver-fox farm with forty-nine breeding vixen, and in addition the farm employs hunters to bring in pelts of squirrel and ermine.

The ranch is an important source of meat for local towns. The smetana and the reindeer, cattle and horse hides and fur pelts enjoy markets further afield. However, the fur farm has not been found economic, since the "lion's share" of the profits go to the city furriers of Leningrad, the "Fur Capital of the World".

Chapter VII

Primitive Nomadic Gatherers and Cultivators and Primitive Sedentary Cultivators

Introduction

THE pre-agricultural gatherer is probably the most primitive of all peoples. The limited climatic and soil tolerance of most plants, as compared with the adaptability of most animals, restricts such people to the very limited geographical environments where plant food is available at all seasons, that is, to the tropics. Although they may acquire plant lore, they do not have to pit their wits against the evasive cunning of wild animals as do the hunters, and so their technology is relatively limited.

Nevertheless, the gatherers enjoy a more orderly social existence than the hunters, and when they acquire the skills of cultivation, potentially they enjoy better opportunities for leisure and the acquisition of civilisation. Cultivation is also economically more productive than pastoralism, since animals are inefficient converters of vegetable matter into food. While primitive hunters tend to achieve an easy level of prosperity, they also tend to stagnate at this level, restricting subsequent change, as do the Masai in East Africa.

In this chapter the Semang gatherers are employed to illustrate the purely gathering economy, and the Boro of West Amazonia to illustrate the primitive nomadic cultivator.

The problem of growing population and attendant land hunger may eventually necessitate a transition from nomadic to sedentary cultivation, but this is no simple transition. In the words of P. H. Nye and D. J. Greenland, *The Soil under Shifting Cultivation*, the "...division between nomadic, semi-permanent or primitive cultivators, is ... blurred". Whereas the "Boro shift their cultivation every few years, the Ashanti in Ghana live in towns or villages that have endured ... for centuries", yet they abandon one clearing in favour of another, and do not come back to it for a generation or so, i.e. they use an extremely long fallow period. Some people abandon land and settlement, others retain their settlement but move their cultivated clearings about within extensive surrounding territories.

More significant to agricultural studies, however, are the truly sedentary cultivators, who have learned in any of several ways to maintain soil fertility under continuous cultivation. Such peoples include the Anglo-Saxons, who developed rotational techniques which in evolved forms are still employed today in the agriculture of the British Isles and Western Europe. Irrigational techniques and very extensive cultivation enable the Hopi Indians of the Mesa-lands east of the Little Colorado Valley in the Far West U.S.A. to maintain soil fertility under continuous maize-cropping. Intensive peasant subsistence cultivators in South-east Asia maintain remarkable levels of productivity on dangerously overcropped holdings by intensive applications of organic fertilisers and techniques of cultivation other than rotation. Scientific technology has supplanted many of these primitive solutions to sedentary cultivation, or is in process of so doing for an increasing number of peoples, but primitive systems of this type still support many people over large areas of the earth's surface.

The Semang Gatherers of North-east Malaysia

The Malay Peninsula was at one time an equatorial forest region with mangrove swamp-fringed shores and an interior of dense forests traversed by great sluggish rivers. These forests only thinned out where temperatures were reduced by altitude over the central highlands. Gradually, over a period of hundreds of years, however, Chinese immigrants, and more recently Indians, Dutch and English, along with coastal lowland Malaysians, converts to the Muslim faith, have made incursions even into the more remote parts of the northern interior, exploiting tin and timber, and developing agricultural production. The aboriginal population, notably the Semang and Sakai, survive in the most remote hill-forest lands of this northern interior. The regions occupied by the Semang are shown in Fig. 55.

The environment

The tropical rain forest consists of giant timber trees such as teak, standing 30–40 m high, with tree foliage restricted to a top canopy which excludes all but a filtered green light from the forest floor. At about 10 m above the floor a parasitical growth of creepers is suspended from the giant tree-trunks, covered in bright flowers and providing concealment for brightly coloured birds and great coiled pythons. The massive trunks of such trees are often of a buttressed type, and grow out of bare clay soil, since the vegetation below is restricted by their root systems, as well as by the shade they afford. There are, however, dense patches of undergrowth wherever the great trees permit, and these include tall thickets of bamboo.

Numerous gullies and sluggish tributary streams join up to form massive swollen muddy rivers which carry prodigious quantities of alluvium suspended in the floods brought about by the excessive equatorial rainfall of the area, which also experiences a monsoonal element in its late summer maximum. Rain tends to come in violent electric storms

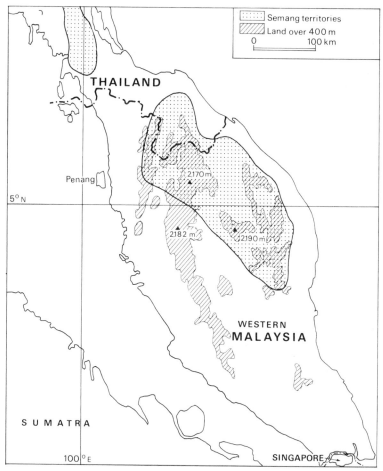

FIG. 55.—*Territories frequented by the Semang food gatherers.*

accompanied by turbulent winds and sheet lightning, which occur in late afternoon, and total rainfall for the year is something in excess of 2,000 mm. Temperatures likewise are usually around 27°–30°C for much of the year, although there is a slight summer maximum. On the higher land, lower temperatures of 15°–18°C may occur, with nightly temperatures falling considerably, being chilly rather than balmy and warm. The profuse vegetation is matched by an abundance of wildlife: small monkeys, wild pig, the orang-utan ("wild man of the woods"), tigers, panthers, leopards and elephants.

The Semang people

Among these tiny Negritos even the men rarely reach 1.5 m in height. They have broad noses, dark brown skins, thick everted lips and tufty hair like Negroes; hence the term "Negrito", which might be translated as "mini-Negro".

A group rarely exceeds thirty in number, and is comprised of the "old man", the oldest able-bodied male, his wife and their children—possibly including sons with their wives and children in the case of a mature group. On reaching puberty a boy, evicted from the group, will eventually select a girl in another group, which he will join for several years in order to cohabit with her. When his bride has borne him one or more children, he will take his new family back to live with his own group. Thus sporadic contact is maintained between contiguous groups, but dialect differences increase within short distances, and next but one groups may be almost unintelligible to each other. Thus, as noted in the description of the bushmen, isolation accounts both for stability and for slowness of innovation among such "primitives".

The group's traditional territory occupies some 50 km². An overall density of about 1 person per 2 km² is fairly typical of forest gatherers. Within this country the group migrates methodically under the direction of the old man in search of sustenance. Fruit within the territory belongs exclusively to the group, but gathering may be conducted within the territory of another group, and pursuit of a wounded animal may also involve incursions upon the lands of the neighbouring peoples. Two features of Semang life are particularly relevant to this study; techniques of gathering, hunting and fishing, and a technology based upon available raw materials, particularly upon bamboo which provides a remarkably versatile and easily-worked material.

Techniques of gathering, hunting and fishing. A group must move at least every three or four days so as not to exhaust the resources of a particular locality. This also takes advantage of the fact that, in the absence of seasonal change, all localities have a fairly constant number of plants reaching fruition at any given time, which makes it practical to secure a regular food supply by means of perambulation.

Each morning women and their children, often accompanied by their menfolk, scatter with their fire-hardened digging-sticks in search of berries, nuts, pith, leaves, shoots, roots and tubers which they gather and carry in matwork baskets slung on their backs. Even young children contribute in this way. Foraging is a daily task, since food cannot be kept long in the hot, humid conditions that prevail. Snacks are taken throughout the day, but all meet at night for a communal evening meal. Fruit trees such as durian are usually harvested systematically by the group. While women concentrate almost exclusively upon gathering with a little fishing, men hunt or fish when opportunity offers, but gather alongside their womenfolk at other times.

Hunting for the Semang is limited to small creatures such as rats, squirrels, birds and lizards, or occasionally wild pigs or monkeys. Their main weapon is a pliable wooden bow strung with sinew or bark fibre. The arrows are smeared with ipoh paste, the sap of the upas tree which is instantaneously fatal to small animals. Small game may be snared with noose or spring-trap, and birds may be trapped using splinters of bamboo smeared with juice from the wild fig trees, which adheres to their feet. The Semang sometimes adopt the bamboo blow-pipes developed by the Sakai and Savage Malays.

Women may join the men in fishing when the group reaches a large river. Small fry may be scooped out with bamboo scoops, while large fish are more often gaffed with a pointed palm-leaf stem. "Beaters" may drive the fish upstream towards the spearmen. The Boro's skill of poisoning the water is unknown to them.

Technology. Bamboo, the stalks of giant grasses, is by far the most significant of plants to the Semang. It is pliant and tough, and its tubular form makes it versatile of application. Bamboo is employed to fashion fish-scoops, quivers, cooking-pots, knives, hair-combs, and various other implements and utensils. Fire-hardened bamboo can carry an edge sharp enough to slice fish, meat and green bamboo.

Shelters constructed for the brief sojourn in any locality consist of a rattan-thatch wall woven horizontally between several poles, capable of being pegged down lower to afford shelter against torrential rains, or released higher to serve as a windbreak. A sort of raised bamboo couch enables members of the family to sleep out of contact with damp soil or vegetation. Fire is generated by friction, but to avoid this irksome task in the humid conditions that prevail, torches of bamboo or "ropes" of damar resin wrapped in dry leaves are used to retain or transport fire as required. Clothes are confined to girdles of beaten barkcloth, for which the technique of manufacture has probably been acquired recently from the Savage Malays.

General Introduction to Bush Fallowing

It has been from such humble origins as those described above, with people wholly dependent upon their immediate environment for survival, that cultivators have originated. Indeed, the bush fallowers are found in similar tropical forest environments, where the sheer prodigality of nature is sufficient to sustain people who lack the most elementary of cultivation skills.

Bush fallowing, or primitive nomadic agriculture, is largely confined to the world's tropical forest regions, where it has been given such names as "Milpa" in Central America, "Roça" in Brazil, "Conuco" in Venezuela, "Masole" in the Congo region, "Ladang" in Malaysia and Indonesia, "Jhum" in India, "Caingin" in The Philippines and "Ray" in Vietnam.

General characteristics include:

1. the selection of a site which is then cleared by girdling and burning the great trees, earning it the name of "slash and burn cultivation";
2. the planting, in individual plots, of cereals (maize in the Americas, millet in Africa, and rice in Asia), root tubers, a variety of vegetables, and sometimes such small "cash-crops" as sugar-cane or cotton;
3. the keeping of a few poultry, pigs or goats;
4. supplementary hunting, fishing or collecting;
5. finally the abandonment of a settlement after a few years as a result of one or several of the factors of soil exhaustion, jungle encroachment, disrepair of the village, or the ravages of disease, or of pests such as rats or termites.

The Boro, Nomadic Farmers of West Amazonia

The environment

The Boro occupy about 13,000 km² of dense equatorial forest in the Upper Amazon Basin, a region traversing the boundaries of Brazil, Peru and Colombia, and bounded on the north by the River Japura, and on the south by the River Putamayo or Issa, both tributaries of the River Amazon (*see* Fig. 56). The Upper Amazon and its tributaries are vast and sluggish (the Issa is already about 200 m wide in the Boro Territory) and the vast quantities of alluvium they have brought down

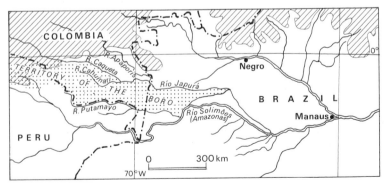

FIG. 56.—*Environment of the Boro of the Upper Amazon Basin.*

and deposited in flooding, rich in humus, have given rise to a soft black soil.

The climate of this upper basin displays little variation. Temperatures hover at about 27°–30°C, and likewise rainfall is heavy all year round, though heaviest of all around November and May. In the prevailing humidity, rivers are liable to flood without warning, e.g. the Cahuinari, tributary of the River Japura, has been know to rise 7 m in a single day.

In general terms, the vegetation is very like that described earlier in this chapter with reference to North-east Malaysia. The only effective communications throughout the region are the great rivers, which, by breaking the tree canopy and affording a view of the sky, make possible easy movement over long distances.

The Boro people

There are an estimated 10,000 Boro, living in about fifty independent settlements, but speaking closely related dialects of a single language. They are brown skinned, and characteristically Mongoloid.

A Boro settlement may consist of one, occasionally two, large communal houses, situated near to an open site reserved for ceremonial dancing and close to their cultivated ground. (A large settlement may be made up of several houses, but this is rare.) Each house is about 20–23 m square and about 10 m high, and provides accommodation for more than fifty people. The corners of the house are rounded, and the ridge roof extends nearly to the floor, overhanging the low vertical walls. The entire roof and walls are constructed of a kind of thatch made of folded palm-leaves set in split bamboos, each portion tied to the framework and overlapped with the next. Sometimes the vertical walls are covered with matting. Inside, these tall airy buildings are reminiscent of a circus marquee. Each family has its own fireplace against the wall where its personal property is stowed. Settlements are always kept well away from the main rivers to minimise the risks of flooding and enemy attacks by water, and to avoid the insects. The path to the river is deliberately made difficult to follow in order to fox enemies. The houses are rarely repaired, and when the thatch has succumbed to heat and damp, fertility of the plots is declining, weeds are encroaching, or enemies have begun to find their way to the village, a new village and clearing are established, only a few kilometres away.

Apart from some intermarriage and similarity of language with nearby settlements, there is no sense of political or social unity between them. Each settlement will have a chief who organises its defence and welfare, and whose household is enlarged by the inclusion of women taken captive in war, and of youths who contribute to his prosperity. There is also a medicine man who effects faith cures, and identifies the causes of any misfortunes which overtake the group. He carries a miscellany of charms in a skin bag, including a rattle, a giant condor claw, a string of quartz beads, and a pearl-shell cup.

The settlement has continuous descent through the male line, that is, it is patrilineal, whereas women leave to marry men from other settlements. All children regard themselves as brothers and sisters and all adults as relatives. This sense of solidarity is reinforced by the clan name, which is transmitted to all children born in the community. There is also a lively tradition of stories and wise sayings.

A peculiar custom is that of "couvade", in which it is taboo for a woman to eat animal food when pregnant, yet after birth the husband assumes this taboo; and whereas the wife may work the day following the birth, the husband may take to his bed for a month, and is forbidden to make or even touch weapons or tools in this period.

Their crafts and technical skills, apart from house building or pottery, are little more advanced than those of the gatherers. It seems that cultural rather than physical restraints have retarded South American peoples more than their African counterparts, so that in place of active trade there is an atmosphere of intertribal isolation and distrust.

The women make coil pottery from clay, which is baked under the fire. Some large receptacles are made of bark or wood. The women also make netted hammocks out of fibre cords, leg and arm bands, and matting, but they have not acquired the art of true weaving. Both sexes produce very effective basketry from cane, bark and palm-leaf, to make the large decorative carrying baskets used by women for bringing home their produce, or by men for bringing home game, and which is suspended by a headband in order to free the hands.

The Boro have learned to remove body hair by spreading the skin with latex, which when dry can be pulled off with the hair. Each house has a pair of large signal slit-gongs made of hollowed-out hardwood tree-trunks—a small high-pitched "male" gong and a larger "female" gong. The drumsticks used to play the two gongs are also headed with latex. Some smaller monkey-skin drums, along with flutes and pan-pipes, are used at feasts and festivals.

Dug-out canoes for river travel may be laboriously hacked and burned out of 6–7 m logs. A settlement may keep two such boats for use during occasional feasts, for barter with a friendly neighbouring settlement or for sending out war-parties.

Clothing for men is made of bark cloth, stripped from the tree, soaked and beaten. Women go naked. Both men and women use wood and shell plugs to elongate their earlobes and lips.

As regards their economy, the Boro depend upon cultivation of their clearings, with subsidiary hunting, fishing and gathering. They keep no domesticated animals. Apart from a large clearing reserved to the chief, most other cultivated clearings are scattered over an area of some kilometres, the furthest out being sufficiently far away to require the owner to construct a temporary hut for overnight stays during the planting season.

When a new clearing is made, at the end of the wet season, the clearing party girdle the trees with their stone axes. These axes have a square blade of polished stone, grooved to retain the binding which secures head to handle. They have to be traded because there is no stone locally. When the large trees are dead they are felled by burning, and the undergrowth is hacked away. The tree stumps that remain soon decay under the prevalent hot wet conditions and the onslaught of ants. The men then rough-dig the ground, but subsequent cultivation is undertaken by the women, who use a digging-stick. Planting goes on continuously to ensure a steady flow of fresh food, although the main harvest comes in the heavy rain season.

After two or three harvests the soil is exhausted, after which it is not tilled again although it may be revisited in search of self-plants or the survivors of former cultivated crops. The main crop is manioc, from the roots of which cassava is prepared as described below. The cultivated variety of manioc has larger tubers than its wild counterpart. The first crop is harvested after eight months, but some kinds grow for two or three seasons. New plants are propagated from cuttings from the old growth, each in a separate hole. Other root crops include yams, and sweet potatoes, while pumpkins, peppers, beans, pineapples and a few fruit-bearing trees are also grown. Maize is very little grown.

Next to manioc the most important crops are coca and tobacco. The seeds of coca are planted at the start of the heavy rain. A few will germinate and grow into a bush 2 m tall, which will come into fruit after eighteen months and continue to bear fruit for a further thirty to forty years. Only the men tend and plant coca, and the drug is only permitted to men. Similarly, although the women plant and prepare tobacco, only the men use it—to lick, not to smoke. The coca contains cocaine, which is made up into a powder with lime, baked clay and cassava flour, and which, when shot into the mouth, enables the addict to go for several days without sleep, food or drink.

Tobacco is prepared in paste form with cassava starch, and stored in a small nut-shell pot. When men meet to settle disputes or public affairs, or wish to bind agreements, they sit around dipping their sticks into the pot and licking the end. Agreement may be indicated by accepting a dip when the pot is ceremonially passed round.

When the manioc root is 0.25–1 kg in weight, at two seasons old, it is ready to be dug for food as required. First it is soaked to loosen some of the poisons. It is then grated on an oval wooden board and packed into a "squeezer", a long tubular structure of plaited palm-bark some 2.5 m long, which is suspended from a rafter, and to the bottom end of which is fixed a long horizontal pole. A woman sits on the end of this pole, bearing down with all her weight, so that the contents of the "squeezer" are compressed and most of the juices pressed out. The pulp can then be dried and powdered into flour, which is heated to expel further gaseous poisons. Finally the flour is kneaded and fashioned into loaves of unleavened bread.

The poisons expelled from the manioc are not wasted; boiled into a paste and flavoured with pepper and fish, they provide a sauce for the cassava. Manioc leaves are also boiled and eaten as a vegetable.

The main meal for the Boro is in the evening, after the day's game has been brought home. Each family dines from its own stock-pot on a stew of grubs, offal, fish and larger game, flavoured with peppers and eaten with cassava bread. The chief's cooking-pot is replenished by all the unattached boys in his retinue, and his wife and female dependants provide the cassava bread for all who feed at his hearth.

Supplementary foodstuffs are afforded by hunting, fishing and gathering. The men hunt tapir, peccary (bush-pig), ant-bear, sloth, small monkeys, rodents, birds and reptiles. Hunting weapons and devices employed include light spears and blow-pipes. The latter are 2.5–4.5 m long, about 6 mm bore, and accurate up to 17 m, but with a range of up to 50 m. They are constructed of reeds, and used to fire a poison dart with a notched end designed to break off in the wound leaving the poison in the victim's body. The Boro hunter carries his poison in a tiny pot slung around his neck. An important ingredient in Boro poisons is putrifying animal matter, but for more effective poisons the hunters trade with the Andoke, who use the sap of the *Strychnos toxifera* in the production of the lethal paralysing poison "curare".

The Boro are not interested in fishing techniques involving the use of baited lines, nets, traps or spears, but prefer to drug the fish in a section of river, using pounded babasco root. The dead and stupefied fish are then gathered at a point downstream behind a weir of wattle fencing.

As already indicated, the transition between nomadic cultivators like the Boro and true sedentary cultivators is very ill defined. Clearly it has been necessitated in different regions at different times, as the supply of land has become insufficient to permit easy abandonment of one plot and the cultivation of another. Cultivators have been obliged to conserve fertility by employing solutions, some of which have displayed considerable ingenuity.

Primitive Sedentary Rotational Farming with Fallow Period

When the Anglo-Saxons first took up the occupancy of England it was as immigrants from an expansive region of open country east of the Rivers Weser and Elbe, where the continental-type climate gave rise to predictable hot dry summers suitable for the development of a cereal-based economy with subsidiary livestock.

Archaeological evidence suggests three stages in the Anglo-Saxon settlement of England, undertaken at different times in different parts. These included an entrance phase characterised by "skeletal infiltration", an expansion phase of "colonisation and conquest", and a terminal phase of "static consolidation". In the latter phase, if not before,

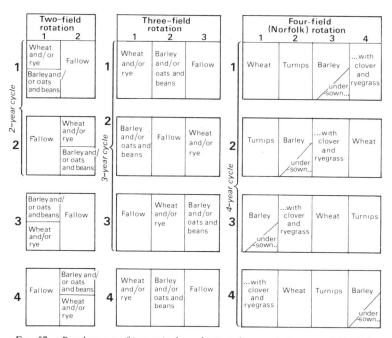

FIG. 57.—*Development of increasingly sophisticated crop rotations*, A.D. 700–1850.

individual leaders carved out agricultural estates and established permanent settlements. In order to conserve soil fertility under a variety of unfamiliar soil conditions, they evolved rotational cultivation with fallow, and in a country where so much land was unsuitable for cultivation, the climate unpredictable, and communications difficult, they gave greater stress to livestock in their economy, thereby ensuring greater security of livelihood and a high degree of local self-sufficiency.

According to *The Open Fields* (C. S. and C. S. Orwin, Oxford University Press, 3rd edition, 1967), they probably practised "an extensive field husbandry.... This evolved (perhaps through a one-field system—early abandoned because it so rapidly exhausted the land) into a two and a three-field system—both of which were in fairly general use in England 900 years ago. The three-field system was an evolution of the two-field system—proceeding variously and carried to different stages of completion in difficult places." In other words, the farmers cropped as intensively as they found their land capable of sustaining it, keeping 50 per cent fallow on poorer land, but only 35 per cent fallow on superior land.

In the seventeenth and eighteenth centuries the new post-enclosure consolidated land-holdings, improved methods of husbandry and the introduction of new crops which made more diversified land-use possible, contributed to the introduction of what came to be known as the Norfolk Rotation (*see* Fig. 57). This four-field rotation was soon adopted widely in other parts of England.

This summary illustrates how one people have moved towards increasingly effective systems of sedentary cultivation over a period of a thousand years. More contemporary developments have included further diversification of crops, the development of effective inorganic fertilisers, and new mechanised farming techniques which have favoured the development of arable farming. The highly sophisticated cropping systems which have evolved are considered in Chapter XII.

Chapter VIII

Commercial Extensive Livestock Farming

Origins of Commercial Extensive Systems

IN the previous chapters we have seen how the mastery of fundamental techniques of domestication and cultivation was achieved, making possible sedentary modes of living and higher levels of civilisation wholly barred to gatherer and hunter. The systems so far considered, which once covered the entire globe, could, however, only provide immediate subsistence to their practitioners. Today they are melting away before the onslaught of the capitalist spirit and the applications of scientific technology, both of which developed initially in Western Europe.

When during the fifteenth century European colonists, urged on by the capitalist spirit, began to open up vast continental territories all over the world, a process continued in the subsequent two centuries, they ruthlessly dispossessed indigenous gatherers, hunters and primitive cultivators in order to impose new and alien systems of agriculture. In the eighteenth and nineteenth centuries the process received considerable stimulus from the Industrial Revolution, which originated in the British Isles and initially spread into Western Europe, bringing about there the evolution of commercial intensive systems of agriculture which will provide the subject of subsequent studies.

In areas outside Western Europe, such as North and South America, Australasia and more recently parts of Africa and Siberia, however, Europeans have been responsible for introducing commercial extensive systems, designed to operate successfully with very few agricultural workers and with the very large tracts of land available. In such temperate and tropical grasslands commercial extensive livestock farming was first introduced, although commercial extensive cereal cultivation and mixed farming have since ousted livestock out of all but the most arid interior regions. In the tropical forest regions commercial plantation agriculture has been developed. An examination of Fig. 20 will illustrate these observations, more fully considered already in Chapter IV.

Diversities within the System

In some regions colonised by the Europeans, the indigenous peoples have been dispossessed of their land and placed upon reservations, assimilated or exterminated, and commercial extensive livestock has developed independent of any previous agricultural practice. However, as already indicated with reference to the nomadic reindeer hunters of north-east Siberia, some primitive nomadic pastoralists have also begun to evolve towards the commercial livestock system under pressures imposed by land shortage, or as a result of contact with more sophisticated people who have either coerced them, or motivated them with their own commercial spirit.

Animal species found most suitable for this system are gregarious herbivores such as sheep, cattle and horses, but others which have been successfully exploited include reindeer, goats, buffalo and camels, and to a lesser degree the yak, llama and alpaca. Since space precludes any attempt at a comprehensive treatment of even the more conventional aspects of the system, such as cattle ranching, fat lamb or wool production, attention is given to illustrating regional diversities as well as the essential unity of the system.

General Characteristics of the System

The main species to undergo domestication have been the larger herbivores, which have proved particularly suitable because they are gregarious, which means they are more easily kept together, while some species have acquired an amenable disposition towards man as a kind of provider-protector. The main purposes for which man utilises such animals are for meat or for draught. In the extensive system described here milking is rarely practical at the large scale of operations employed.

Herd animals are more readily utilised in this system because of their distinctive social organisation in the wild. In general, the animals are born in roughly equal numbers of male and female. To provide for improvement of the genetic stock, however, the social structure of the herd does not permit all animals to procreate. Instead a system of natural selection geared to the survival of the fittest operates. This system, under which only the strongest, most aggressive or most cunning male is able to win access to breeding females, is best adapted to the rigours of the natural environment. A few exceptional males build up a "harem" of up to fifty females, with whom during the period of their prime they reserve exclusive sexual rights. Many more maintain smaller herds, to whom they in turn provide protection and leadership. As soon as the herd leader begins to decline in health or faculties, however, he is rapidly supplanted by a younger or more vigorous male. Young males are accordingly viewed as competitors, and therefore driven out on to the periphery of the herd by their sire the herd leader. Here they engage first in combat with one another, before the best among them is ready to challenge the supremacy of a herd leader for control of his females. So it is that the finest male stock is able to imprint its character upon the entire herd, and those enjoying less than the best of health, stamina and potency forfeit their breeding access to the herd. It is this social system, in which only one male mates with several females, that gives man his chance to manage such species for his own ends.

Under commercial extensive livestock farming men are too few in number to manage the herd from day to day. Instead man intervenes in the social organisation of the herd at this crucial point of breeding selection. In nature, as has been indicated, it is the strongest, healthiest, most virile and most cunning males who reserve the largest retinues of females for their exclusive breeding use for the longest period, and so leave behind them the most progeny. Instead man requires propensities such as the tendency to fatten quickly for good lean meat, or qualities like tractability and strength in draught animals. He has therefore learned to calculate the essential breeding ratio between male and female for any given species so that the total stock does not decline in number, and to select those male animals most likely to transmit the looked-for qualities. Man has taken the process of selection to himself subjecting all males that are surplus to his requirements to an early castration—desexing by removal or destruction of the male gonads—which eliminates them entirely from the competitive urge to breed. Furthermore, the operation changes the entire endocrinal balance of the male, rendering him more rapid in fattening and more docile in draught. Such geldings may, with impunity, be set at liberty with the rest of the herd.

The yearly routine on any such commercial livestock enterprise will be very similar. In spring the young are born and throughout the summer the annual round-ups proceed in various sectors of the estate. Each category of animals within the stock is subjected to the appropriate management procedures outlined below.

1. Aged stock, beyond an acceptable breeding age, are "culled", that is, withdrawn from the herd for subsequent marketing. They may be sold for a few more years' breeding under less stringent conditions elsewhere, but more usually they are gelded or spayed preliminary to quick fattening for early slaughter. Such animals provide low-grade meat, hides or skins, bone-meal, dried blood, animal fats, glycerine, glue and fertilisers.

2. Breeding stock undergo routine veterinary attention such as drenching and dipping to eliminate internal or external parasites or infections, and to identify any animals not desirable for breeding. At the same time the opportunity may be taken to shear animals which produce marketable wool or hair, such as sheep, goats or alpaca.

3. The followers or yearlings include young females who are now ready to be released into the breeding herd as replacements for the aged stock which have been culled as surplus to requirements. They also include last year's geldings or fat-stock which are now

ready to be marketed for "finishing", i.e. final intensive fattening for early slaughter, or to go for draught animals.

4. Finally, this year's "crop" of young animals must be attended to. Most of the young females, which will be retained for herd replacement, are branded or earmarked for permanent identification. A few undesirables may go for fattening. The young males undergo breeding selection. A few of the best males may, on the basis of certain breeding criteria, be retained "entire" to replace aged males culled from the breeding stock. (Alternatively it may be policy to buy in "new blood" from specialised breeders or to employ artificial insemination, in which case no such selection is necessary.) All other young males are castrated along with any renegade males who on a large holding may have evaded last year's round-up, and have currently been apprehended with a following of females. Horned cattle may be de-horned, and sheep docked, at the same time as branding/earmarking and castration, since this may be the only time, on a large holding, that they will be seen for another year. Animals which are not going to be kept for breeding receive only a superficial earmark or brand over which a subsequent owner will be able to superimpose his own mark. These young animals are then released with their dams to mature or fatten until the subsequent round-up.

Upland Sheep Farming in the Harlech Dome of North Wales

The environment

The Harlech Dome is an anticlinal structure of Palaeozoic shales, slates and gritstones bounded on the western, seaward side by an ancient raised cliff-line, and a coastal plain composed of a raised beach fringed with sand-dunes. The dome is bounded on its north side by the Afon (River) Dwyryd, on its south and south-east sides by the Afon Mawddach, and on its north-east side by the Ceunant Llennyrch, a tributary of the Dwyryd, and by Llyn Trawsfynydd. Its maximum north–south extent is about 20 km and east–west about 16 km. The drainage of the Harlech Dome, emanating from the gritstone core of the Rhinog Range, is radial, and among the fast-flowing short rivers are the Afon Crawcwellt on the eastern flanks and the Afon Artro on the western side (*see* Fig. 58).

The interior of the dome has undergone severe glaciation, leaving its uplands bare and bereft of soil. The average annual rainfall at sea-level is about 1,500 mm, rising to over 2,000 mm in the higher areas. This well-distributed rainfall displays a winter cyclonic maximum. Summers are cool, with July average temperatures of about 15°C. January temperatures near to the sea are surprisingly mild, about 4°C, although on high ground they may be around 0°C.

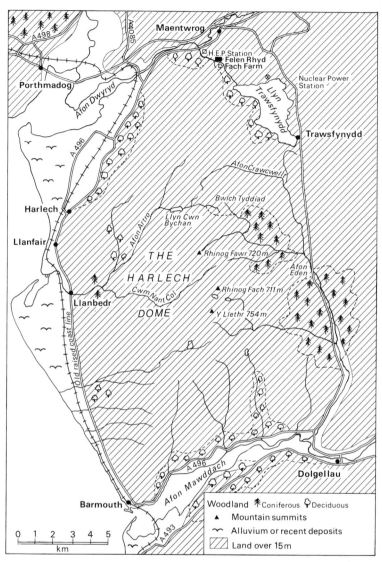

Fig. 58.—*The Harlech Dome.*

The natural vegetation on the uplands was originally birch, oak and
mountain ash or rowan, much of which has now been felled to make
way for sheep-runs, although some deciduous woodland survives along
the steep inaccessible slopes of river valleys. Where the forests have
been cleared, short well-grazed mountain grass or bracken and heather

occur; on bare exposed rocks, lichens and mosses, and on undrained hollows, cotton-grass, bog-bean and sphagnum moss may be seen. Along the flat alluvial valley floors and up the adjacent hill slopes, there are small, irregularly shaped fields of improved pasture, bounded by dry-stone walls. These walls are often composed of erratics gathered up when the fields were de-stoned. The long, low-built farmhouses are usually constructed of local gritstone and slate quarried on the nearby hills.

The slopes of several principal valleys and many mountain flanks are planted in conifers, as a result of the reafforestation policies of the Forestry Commission, who take over the land of farms abandoned as uneconomic by the dwindling and ageing farm population. Surviving farms are often amalgams of many smaller, older holdings, which together provide units of economic size in the competitive climate of the European Economic Community.

Felen Rhyd Fach Farm, Maentwrog

This farm, near Harlech in North Wales, occupies a large, irregularly shaped holding of some 135 ha, one which only a few years ago was occupied by seven farms and supported twenty-seven people, yet now supports a family of five, such has been the pressure for amalgamation into larger more economic units (*see* Fig. 59). Mr. Gwynedd Pritchard has farmed here since 1954. The land is two-thirds owned by him and one-third tenanted by him.

The name of the farm means "small yellow ford", referring to its situation near to a fording place on the Afon Dwyryd. According to a folk tradition embodied in the Welsh *Mabinogion*, Prince Pryderi of Dyfed was slain at Felen Rhyd and buried at nearby Maentwrog.

The farmhouse and outbuildings are very old and built of local slate and gritstone, with the exception of one large modern building. They stand at about 32 m above sea-level.

The land. The lowland part of the farm is composed of clay, but is in part overlain by alluvium along the floodplain of the tidal Afon Dwyryd, where embankments about 3.5 m high were constructed some 160 years ago by the Oakleys, the principal Blaenau Ffestiniog slate-quarry owners, in order to reduce flooding. Still lower down the estuary the clay is overlain by peat. Because of a very high water-table, the clay tends to become waterlogged. To combat this tendency some 26 ha have been drained at the cost of about £600 per hectare. These water-pastures were originally covered in rushes. The adjacent hill land, which rises to 212 m, is based upon local gritstones and shales. The thin acid shaly soil originally supported a natural woodland of oak, birch and mountain ash or rowan, with bracken, gorse and heather on the more exposed parts. Most of the woodland was felled in remote times to make

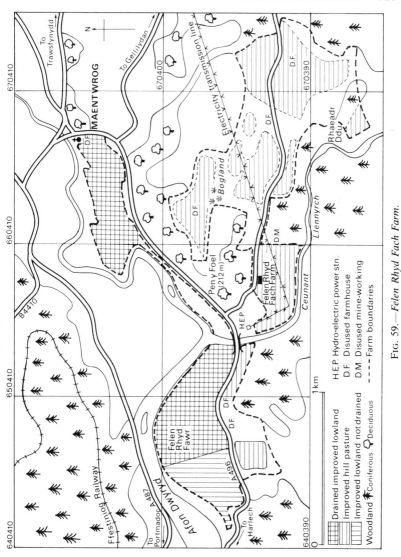

FIG. 59.—*Felen Rhyd Fach Farm.*

way for sheep-runs, but it survives in patches on the less-accessible slopes. Where possible Mr. Pritchard has replaced the original vegetation on lower hill slopes and on the lowlands with improved pasture, which includes perennial ryegrass and white clover. Nevertheless, such improvement is a never-ending struggle, since, despite expensive ploughing, seeding and liming, as well as carefully controlled grazing, such pasture constantly threatens to revert to type, with the less vigorous grasses being ousted by more resilient bracken and gorse. The farm boundaries touch upon several small stands of commercial timber, conifers such as Sitka spruce, and Douglas fir. There is evidence in one place of a small exploratory mine adit where quantities of galena-lead ore and chalcopyrites—copper pyrites—were found in association with quartz.

The drained lowlands provide the most valuable land resource on the farm, for it is here that cattle are pastured in late spring; where hay and silage for winter fodder are produced in early summer after the cattle have been sent up the hills in order to permit the grass to grow; and where the ewe lambs are "wintered" in their first year, thereby obviating the need, as in past times, of wintering them on a lowland farm from 15th October to 5th April, and so incurring considerable expense.

In recent years many dry-stone walls which once separated the tiny lowland fields have been removed, for example seven such fields are now incorporated into a single 5 ha field. The main reason for such removal has been that the walls were rather low, just adequate to confine cattle but not sheep, which tended to jump over them anywhere, scattering stones and leaving gaps. They were untidy, bracken and brambles sprouted at the bottom, and their upkeep involved considerable time and cost. The new larger field lends itself to mechanised fieldwork, and an area of about 600 m² was actually released when about 600 m of wall was removed.

Crops. No crops are grown on this farm apart from grass, for which the relatively cool wet summers lend themselves. This is not because cropping is impractical—indeed potatoes, oats, even barley, swedes, kale and rape are often grown, and have been traditional, in this area, despite risks of flooding in lowlands and the prevalent stony hill slopes. The real reasons are economic rather than physical, for crops are more labour-intensive, or require additional specialised machinery, while grass provides much needed winter fodder without these additional costs.

Livestock. Livestock and animal products constitute approximately 65 per cent of total gross receipts for the farm, the rest being made up from grants and subsidies, 20 per cent, and tourism, to be considered subsequently, which provides the final 15 per cent.

The main form of livestock is a flock of 480 Welsh Mountain ewes, plus their lambs, which, allowing for some multiple births, may total 540 in a normal year. Each year the season opens with the lambing, which may last from 20th March until the end of April. After the lambing the ewes are released up on to the hills in May. The yearlings are sheared early in June, and the main flock is usually sheared one day about 20th June, when six helpers work to ensure the shearing contractors are kept fully occupied. A single fleece may weigh 1.5–1.75 kg, and the wool crop is marketed in nearby Porthmadog. At shearing time, to avoid mustering the sheep twice, the lambs are dosed and dipped to kill lice, keds and ticks, as well as to prevent fly-strike, the result of which can be an infestation of maggots which will play havoc by burrowing into the sheep's backs.

From late July onwards fat lambs are marketed in successive batches as they reach 27–30 kg liveweight, or an estimated deadweight of 12.5–14 kg. They are marketed through Dolgellau Fatstock Market or through the Welsh Quality Lamb Group, a local market co-operative. It is a remarkable demonstration of the economic performance of the Welsh Mountain ewe that she can, on poor pasture, rear a lamb nearly her own weight in four to five months, especially if crossed with a heavier ram like a Suffolk or Border Leicester. The last fat lambs are usually marketed about Christmas time.

The flock of 480 breeding ewes is made up of successive generations of ewes at two, three, four, five and six years of age. In addition, there are about 110 ewe-lambs selected from the current lamb crop, which will have their own first lambs at two years of age. The ewe-lambs having been retained for flock replacement, older ewes of between $5\frac{1}{2}$ and $6\frac{1}{2}$ years of age are marketed each September. They go to a lowland farmer, for example one on the Lleyn Peninsula, where under the less-exacting conditions prevailing there they can rear lambs for a further two seasons. In September the entire flock undergoes a further mandatory dipping against sheep-scab.

On 20th October the rams are let out with the flock on the mountains, having previously been segregated from the ewes in order to delay lambing until the worst of the winter is over, since a ewe's gestation averages about five months. Some ten rams are kept on the farm, that is, a little above the ratio of one ram to fifty ewes. Between Christmas and early January the sheep are brought down near to the farm in anticipation of the lambing season to come. Here, if the weather is bad, they can be hand-fed on concentrates or hay.

The lamb crop provides 35 per cent of total gross receipts, sheep and ewes a further 5 per cent and the wool crop 5 per cent.

Mr. Pritchard also keeps about thirty Welsh Black cows and a bull, mostly pedigree stock, under the herd-name "Pryderi"—an appropriate name both from the situation of the farm and because this native Welsh breed probably dates from the epic days of the *Mabinogion*. Cattle make up another 20 per cent of total gross receipts.

By late May the calves have usually all been born, and the strongest may be sold as "suckler calves" in the following October, while the late-born calves too small for this market have to be wintered on the farm and fed on expensive concentrates until the following year, a procedure which does not always pay. Nevertheless, Welsh Black cattle, like Welsh Mountain sheep, display a remarkable ability to prosper under stringent conditions, fully justifying the perpetuation of the breed. The cattle spend late spring on the lowlands but are sent to the uplands in early summer. This is transhumance in the infield-outfield tradition, or what is known here as *Hendre*, "winter abode" or lowland, and *Hafod*, "summer abode" or hill land. (*See also* Fig. 6 in Chapter I, and Fig. 92 in Chapter XII.)

At Felen Rhyd the two are comparatively close together, but today this may not necessarily be so, indeed the same principle may be made to operate with lands far apart. Thus, for example, many Merioneth farmers today have lowland farms on the Lleyn Peninsula and Anglesey, and vice versa.

Transhumance is practised to make possible sufficient growth of the lowland pasture for hay and silage-making later in the summer. Some cattle are wintered out of doors, while others are "yard-fed" near the farm. The yield of hay from about 20 ha of improved lowland pasture has been about 5,000 bales per annum in recent years, thanks to resowing, drainage, and also constant liming and the use of artificial fertilisers containing nitrogen, phosphates and potash which are essential to the maintenance of fertility in the face of excessive leaching and "run-off" experienced in such a high rainfall region. Notwithstanding these efforts there are recurrent problems of mineral deficiencies: of cobalt, magnesium and calcium which affect lactating animals, and of copper which, although usually available in the soil, is "locked up" in combination with a high molybdenum content, particularly lower down the estuary. This causes "sway-back" (paralysis) in young lambs, and unthriftiness (slow liveweight gain and poor condition) in young cattle. The greater the grass yield, the denser the livestock may be grazed and the more mineral deficiencies are likely to be felt. Too much lime may also aggravate this problem.

Other problems include the farm's tidal river boundary over which cattle and sheep may wade or swim, and difficulties of access on a farm which extends so far east and west, and is also bisected by the main Harlech road.

Nevertheless, the holding is successfully operated as a family farm with only casual or seasonal labour provided by reciprocal understandings with neighbours, or, as with shearing, by engaging contractors as required.

Tourism. Reference was made earlier to the contribution of tourism to farm income—an estimated 15 per cent of gross receipts. This is made up of earnings from the provision of bed and breakfast and evening meals to farmhouse visitors, from a small caravan park, and from pony-trekking amenities. More relevant to agricultural studies is the developing educational services provided here for higher education establishments and schools, probably the first time such amenities have been provided on an ordinary working farm. They include provision of a lecture room near to the farmhouse, where reference books and display materials can be made available to students, along with opportunities, subject to prior arrangement, for live lectures, a conducted walk around the immediate farm, and demonstrations of sheep-dog working or some comparable seasonal activity. There is also a Farm and Landscape Trail laid out, with commentary provided, by the Snowdonia National Park Authority.

NOTE: After the above study was written with his active help and co-operation, Gwynedd Pritchard was tragically killed in a ploughing accident. The author initially met him as a result of a student visit, and found in him not just a progressive farmer and a champion of education, but a friend. It is now hoped that this study of his farm will stand as a tribute to his life's work in agriculture.

Horse, Cattle and Tobacco Farming in the Blue Grass Region

Kentucky, transitional between East and Midwest, South and North U.S.A., is a state with considerable physical and cultural diversity. In size Kentucky, occupying 104,623 km² ranks thirty-seventh in the Union. The principal crop is tobacco, in the production of which it ranks second only to North Carolina. In 1975, of a total tobacco production for the U.S.A. of 961 million kg, Kentucky furnished 198 million kg. Other important crops include corn (maize), soya beans, wheat and fruit. More than half of Kentucky's agricultural wealth is derived, however, from livestock, principally cattle, horses and pigs. The economic importance of livestock is particularly evident in the Blue Grass region, even though very few farms are monocultural as are traditional cattle ranches or sheep stations. While livestock is the primary interest of the farmers described below, a substantial part of their income is derived from commercial tobacco, other crops or non-agricultural sources.

The Kentucky Blue Grass region

The region occupies 21,000 km² of north-central Kentucky—a gently undulating plateau standing at about 300 m (see Fig. 60). The entire region is underlain at varying depths by mammoth cave limestone, sometimes referred to as Kentucky marble. Above this is a stratum of softer granular limestone, which decomposes to form tricalcium phosphate. Above is a clay subsoil, while the surface is of maury silt—a phosphatic limestone soil, light, fluffy, reddish in colour, and rich in calcium and phosphorus in its natural state. The dense underlying limestone provides a high water-table, rendering this otherwise porous soil sufficiently water-retentive to provide excellent pasturage for livestock,

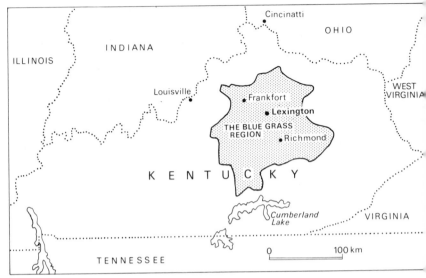

FIG. 60.—*The Blue Grass region of Kentucky.*

yet afford good natural drainage. The soils are naturally rich in minerals necessary to feed the blood, teeth and bones of great herds of livestock. Today, however, the valuable grasslands tend to be overgrazed, exhausting the natural supply of minerals, which therefore have to be supplemented by the farmers themselves (*see* p. 133).

The climate may be described as humid subtropical, and may be illustrated by reference to Lexington, which is situated at lat. 38° N. and long. 84° 36′ W. and at an altitude of 298 m.

The temperatures experienced are as follows:

January mean	1.4°C
July mean	25.2°C
Annual average	13.1°C
Annual temperature range	23.8°C

Rainfall is as follows:

Winter (Dec.–Feb.)	300 mm
Spring (Mar.–May)	321 mm
Summer (June–Aug.)	303 mm
Autumn (Sept.–Nov.)	213 mm
Total annual rainfall	1,137 mm

The early development of Kentucky

Before the Red Indian tribes of the area were defeated at Point Pleasant, West Virginia, on 10th October 1774 there was little or no settlement in what is now known as Kentucky. Daniel Boone established Boons-boro in April 1775, and this was one of the earliest permanent settlements. During the Civil War (1863–6) the Blue Grass tobacco farmers and the cotton farmers of the "Purchase"—a continuation of the cotton-growing region of Tennessee—were slave-owners, and so chose to support the Southern Confederacy. In contrast the farmers in other parts supported the Northern Unionists. Family feuds acquired about this time have persisted down to living memory. These two facts, i.e. later settlement than the East and the rift between supporters of South and North in the Civil War, serve to underline the transitional character of the state.

Blue Grass agriculture today

Chief crops include tobacco, in rotation with sown grass for hay, wheat and corn (maize). Livestock includes the rearing of various breeds of horse, cattle, hogs and, to a lesser degree, sheep. The scale of operations ranges from small tobacco plots or market-garden holdings on a share-crop basis, up to large-scale rearing of thoroughbred horses with tobacco production. Economic pressures have encouraged the farmers to buy non-contiguous holdings, some of which may be let out to tenants, while others are operated from the home farm. Arable and livestock farms may operate as separate farm units, although their contributions to the over-all operation are complementary. Such cases are noted in the studies which follow, although only a single farm unit within the group is described. Several larger farmers also have economic interests outside of, but relating to, farming, for example banking, or the warehousing or marketing of tobacco, such is the present-day complexity of the agrarian economy of one of the U.S.A.'s richest agricultural regions.

Horse rearing in the Blue Grass. Three main kinds of horse are reared. Most valuable of these are *thoroughbreds*, which originated from the Near East, but had already undergone development in England before their introduction by English settlers in Kentucky. A thoroughbred may be reared by the farmer to be sold to a racing stable, or he may race the horse himself. After training as a yearling a horse will begin racing at about two years of age. Following a successful racing career a horse may earn huge stud fees for its owner.

Standardbreds, like thoroughbreds, have descended from Near Eastern stock, but have been bred for "trotting", i.e. harness racing. The fastest racing track in the world, where the famous "Hamiltonian" is run, is at Lexington (Pop. 120,000), which is known as the "Horse Centre of America".

Saddlebreds, which also contain thoroughbred blood, have been selectively bred to provide sure-footed, sturdy riding horses, with a good style and a turn for speed. Today they are show animals, competing in either riding or harness classes. Other breeds of horse reared in Kentucky include the Tennessee Walking Horse and the Quarter Horse.

Tobacco production and marketing. Originally many small subsistence farms in Kentucky also grew a small cash-crop of tobacco, in addition to food crops of corn, wheat and vegetables. Then for a brief phase hemp, for rope, became the chief commercial crop. However, the Federal anti-trust legislation of 1910, by imposing rigorous quotas upon farmers to regulate the area of land which could be used for tobacco, secured the producers of Kentucky, Virginia and Maryland against overproduction, and helped make tobacco a profitable crop by curbing the monopolistic abuses of the American Tobacco Corporation. At a later time, when yields per hectare began to rise substantially, the quota was redirected against output rather than the area of land under tobacco. Each farm-holding has a production quota, but when the land is sold the quota is not automatically transferable, thereby preventing the purchase of land simply for the tobacco-growing rights. Blue Grass land is exorbitantly priced, thanks to its quality and potential. To prevent its purchase by developers, state law requires a minimum purchase of 10 acres (just over 4 ha) per house, too costly for many house developments.

The tobacco grown here is fine-quality "Burley Tobacco" for cigarettes, renowned for its splendid flavour. The Burley belt extends over central Kentucky, southern Ohio, Indiana, west Missouri and part of Tennessee.

Plum Lane Farm

This farm lies about 6 km north-east of Lexington (*see* Fig. 61). The present owner is Clarence Lebus, whose grandfather emigrated from Alsace-Lorraine at about the time of the Franco-Prussian War of 1870. His son, having at one stage defended his interests as a "night-rider", made good and became a man of property. In 1928 he was able to pass on thirty-eight farms to his son, the present owner. Ninata Farm ("Ninata" is Japanese for sunshine) was renamed Plum Lane after Pearl Harbor in 1941, when there was a revulsion of feeling against the name. The new name was inspired by a long avenue of plum trees leading to the farmhouse. This farm occupies 162 ha of maury silt soil overlying the limestone to a depth of about 0.5–6.0 m. Tobacco replaced hemp as the main crop in the time of Mr. Lebus's father, in an era when the farm was the scene of many private race-meetings.

Mr. Lebus's many business interests include the marketing of tobacco, and it was only a few years ago that he relinquished ownership of Lexington's largest tobacco warehouse. In addition to tobacco there are 180 thoroughbred horses on the farm, including 135 brood-mares

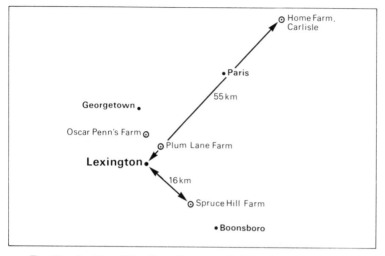

Fig. 61.—*Position of Blue Grass farms described in relation to Lexington.*

and foals and yearlings. The year begins with the foaling from January until June. Since all animals registered with the Jockey Club are deemed to have been born on 1st January of that year it is advantageous for foals to be born early in the year. In March soil dressings are applied on the basis of periodic soil tests, as, for example, when slight calcium deficiency was detected in one field, and a general deficiency in potassium. A good deal of calcium and other elements are absorbed by the sound teeth and dense bones of the animals reared on the land. The pH or acidity tests revealed neutral readings deviating but little from the ideal pH of 6.2.

The last four days of March and the whole of April are given over to ploughing. The corn is set out on 1st May, and tobacco transplanted from the seed-beds on about 15 May. In early August the yearlings are prepared for the sale which takes place on 11th September. Sales for horses of all ages occur during November. Tobacco is harvested in late August and September, racked in a tobacco barn to dry, and finally marketed in Lexington. Corn is used as winter fodder on the farm.

Oscar Penn's farm

Mr. Penn's farm, about 9 km due north of Lexington (*see* Fig. 61), consists of 173 ha of undulating terrain displaying the attractive parkland, tobacco barns and white board fences of the Blue Grass countryside. The attractive large modern house is built in the traditional neo-classical style of many older Kentucky mansions.

Originally the land was part of a 1,500 ha horse-rearing estate held jointly by John D. Rockefeller and his associate, a Mr. Harkness.

Under anti-trust legislation, such large estates were broken up, and this portion passed to Harkness and by succession to his son. Later it was sold to the present owner to finance the development of the main Harkness Estates across the road. The land is composed of maury silt overlying a clay subsoil with limestone at varying depths below. When first taken over, the estate was in pasture, with dense trees, and just two or three small stables. Mr. Penn has cleared many trees and erected a number of modern horse and tobacco barns in addition to his fine

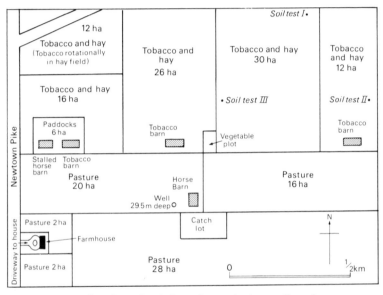

FIG. 62.—*Oscar Penn's horse farm at Lexington, Kentucky.*

house (*see* Fig. 62). Soil tests conducted on the tobacco land in the east of the holding show pH readings of 5.6, 6.2 and 6.8 (the norm being 6.2), calcium content high and potassium in sufficiency. An excellent well 29.5 m deep, with water at 15 m, provides stock with lime-rich water and facilitates the watering of tobacco beds during droughts.

In the early days Mr. Penn kept cattle and sheep, but it was cattle rustlers (65 cattle once vanished in a single night) and ravaging dogs which persuaded him to concentrate on horses alone. Today he keeps 49 thoroughbred mares which at the time of the study had yielded 29 foals, since, unfortunately, virus abortion had taken toll of 44 in-foal mares. There were 34 yearlings currently for sale.

The horse-management year begins with contriving to get mares in-foal as soon as possible, so that after a gestation of $11\frac{1}{2}$ months, they will give birth early in the year, as previously explained. In-foal mares are vaccinated against abortion and all are foaled by June. During August foals are prepared for the yearling sales on 11th September. General horse sales occur in November. Horses are grazed all day and stalled at night.

There is only one commercial crop, tobacco, which is rotated with hay on 16–18 ha of land. The rest of the holding is given over to pasture.

The tobacco season begins with sowing of the seed-beds in March. Planting out takes place in late May and early June. The crop is usually ready for harvest in late August and early September. Harvesting involves hand-cutting with a tomahawk, then six plants are spiked on to a sharp stick which can be suspended on well-ventilated racks in the tobacco barn for curing. The cured leaves are plucked and graded for colour and texture, and made up into bundles of sixty-five leaves to be auctioned to tobacco companies. The tobacco produced here, the Burley variety, is highly regarded for its delectable flavour in cigarettes, while other varieties produced in the area provide burning quality, and yet others are suited to cigar manufacture. Leaves specially grown under cover, and thus free of holes or blemishes, are used as "cigar wrappers", the outside leaves which, if holed, prevent the cigar from drawing properly. Mr. Penn himself owns a 2.8 ha warehouse in Lexington, the nearest tobacco-marketing centre for farmers in Fayette County.

Mr. Penn employs three or four full-time horse workers and a night-watchman. During the peak season when the tobacco is cut, thirty or forty seasonal labourers may be engaged for the harvest.

Mr. Penn has vested the routine management of his farm in the hands of his two sons, both university graduates in business management and agrarian economy. One handles the horse, the other the tobacco and general farming side of the business. The farm is highly mechanised with such machines as a transplanting machine, a high-boy for spraying, a plough, a harrow, a manure-loader and spreader, a hay-cutter and roller, a hay-bailer and loader, and a conveyor to carry hay up into the barn.

Research amenities are provided by the tobacco experimental station and research laboratories at the University of Kentucky, Lexington, as well as the State of Kentucky's diagnostic laboratories.

Home Farm, Carlisle

Mr. William L. Hollar farms a number of non-contiguous holdings totalling 287 ha. He is also a stockyard owner and director of a bank. The Carlisle home farm, about 55 km north-east of Lexington (*see* Fig. 61), is a little outside the true Blue Grass region, but nevertheless enjoys a similar deep water-retentive soil underlain by limestone. However, the land varies in quality, slope and consequential land-use from one section to another. Distinguishable land-use types include lowland with

0.6 m deep loam used as cropland; ridgeland with 0.6 m deep loam also used as cropland; clay soil on undulating topography mainly used as pasture, and which if cropped can only stand one year; and finally, land on steep slopes which has undergone gullying, and which is mostly left in rough woodland and is of recreational value only.

Land-use reflects this diversity. Of the 287 ha, (*see* Fig. 63), about 8 ha are given over to farmhouses and lots, some 6 ha are regularly planted with Burley tobacco, from 8 to 28 ha with hard hybrid corn, and from 2 to 8 ha with soft autumn wheat, while 60 ha are regularly reserved for hay production. The remaining land is given over to pasture.

Livestock comprises 60 Aberdeen Angus cattle and 2 bulls for beef production, and at the time of the study there were 54 calves of 8–10

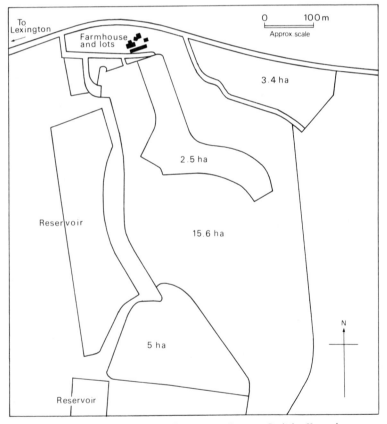

FIG. 63.—*William Lovell Hollar's Home Farm at Carlisle, Kentucky.*

months. It is planned to increase this herd to 100 cows and 3 bulls. Cattle are marketed through Lexington.

Labour includes four men living on outlying sections of the estate who handle the arable farming on a fifty-fifty partnership basis. This kind of agreement, where half the crop is paid to the landlord in lieu of rent, is called "share-cropping". Additional workers may be hired on an hourly basis, or under contract.

Intensive mechanisation includes a hay-baler for stackable rectangular bales, a roll-baler to make cylindrical bales which are left in the fields, a corn-picker, a loading elevator, five tractors and a tractor spray outfit. Technical advisory services available to the farmer include those of the Federal Government, while the University of Kentucky offers facilities for disease research, livestock autopsies and soil testing, as well as courses of training in managerial skills for farmers. Mr. Hollar has himself served on a soil and water conservation board for many years, and has become very conservation conscious. A strong Christian concern with the problem of world hunger has also prompted him to strive towards tripling the food productivity potential of his holding.

Mr. Hollar estimates his agricultural income as derived 62 per cent from tobacco, 34 per cent from cattle and 4 per cent from corn (maize).

Spruce Hill Farm, Athens

This holding, about 16 km south-east of Lexington (*see* Figs. 61 and 64), includes 115 ha of fine farm-land based upon maury silt, and underlain, as previously indicated, by limestone. The soil is naturally rich both in calcium and phosphorus, ensuring an enviable density of bone in the livestock. However, so much calcium is absorbed in this way that over a period of ten years an additional 7,500–10,000 kg per ha must be applied to the land. The limestone actually outcrops in some sections of the farm, and elsewhere lies at about 3 m below the surface. The farmhouse is some 160 years old, one of the first built outside Fort Boonsboro in a period when Athens was called "Boon Station".

The only crop is sown grass—alfalfa, clover and orchard grass—for hay. Mr. Tilson, the proprietor, also owns another farm where the primary concern is with tobacco production, but this is a separate operation, and Mr. Tilson is concerned solely with livestock production. As a one-time agricultural specialist, Mr. Tilson is engaged in genetic research in addition to the farm's commercial operation.

The livestock includes twelve thoroughbred brood-mares, one thoroughbred stallion, and eleven cross-bred (thoroughbred × standard saddle) brood-mares. Following a gestation of $11\frac{1}{2}$ months, foaling occurs at any time between late February and May, since the oestrus cycle in horses is so variable. Whereas thoroughbred markets are exclusively in August–September, the markets for foals, yearlings and "finished hunters" or steeplechasers are all year round.

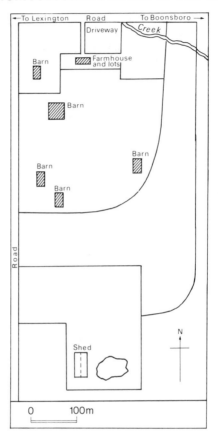

Fig. 64.—*Brian Tilson's Spruce Hill Farm at Athens, Kentucky.*

In addition 40 pure-bred Holstein cows and a pure-bred Aberdeen Angus bull are kept. There are also 20 calves which will be sold as yearlings for beef. The calves are born in one of two seasons, March to May or September to November, and markets are accordingly all year round. Mr. Tilson is researching into improving their disease resistance.

Finally there is a flock of Suffolk ewes, numbering 100 strong, which have been selected from pairs of male–female twins so as to ensure that they originated as two separate eggs. This selection criterion disposes the flock to be of high fecundity. An important quality of Suffolks is their ability to suckle two lambs. The aim is to build up a strain capable of high fecundity, and at present 160 lambs a year are marketed. Lambing is between January and March, and markets for the lambs are in the period May–July, when the lambs are about 40 kg in weight.

One full-time man is employed for the horse management. Mechanisation is directed towards the preparation and distribution of hay. Since the rolled bales, which are stored out of doors, tend to grow mouldy, which renders them unsuitable for sheep or horses although adequate for cattle, much of the hay is block-baled for storage under cover.

Sources of income on this holding include an estimated 60 per cent from the horses, 25 per cent from sheep and 15 per cent from cattle.

Amenities available in this area to help the farmer include the Experimental Research Station of the University of Kentucky, which deals with research into horse diseases, and provides soil analyses. The university also provides courses in agricultural management. In so important a livestock region farmers have recourse to the most distinguished of veterinarians. The State Animal Diagnostic Laboratory at Lexington provides research facilities for livestock other than horses. Finally the U.S. Federal Soil Conservation Service provides invaluable guidance to farmers on the management of their land.

Ranching in Australia

This study serves to demonstrate how, in common with other extensive commercial farming systems, ranching is evolving towards smaller, more intensively cultivated and diversified units. The ranch described is in the Talwood District, near Goondiwindi, Queensland, some 435 km west of Brisbane, and near to the border with New South Wales, at lat. 28° 29′ S. and long. 149° 30′ E. (*see* Fig. 65). Details for the study have been furnished through the kind help of David Pollard, District Adviser to the Beef Husbandry Branch of the Queensland Department of Primary Industries at Goodiwindi.

The holding comprises 4,304 ha and extends roughly 9 km north to south, and 7 km from east to west. The holding is compact, although a main road runs through the property (*see* Fig. 66).

Most of the land is covered in clay soils: ranging from heavy and highly water-retentive to light and moderately water-retentive. There is also an area of grey-brown sand which is prone to wind erosion, and another of red clay loam; both of these are rather less water-retentive. Many of the soils are deep, and the land, lying at about 180 m above sea-level, is very flat. It is therefore well suited to livestock grazing or cereal cultivation.

The average daily maximum temperature in summer (November–March) is 38°C, and there are about 15–20 days when temperatures can be expected to exceed this. Average daily maximum temperatures in winter (May–August), on the other hand, are usually about 7°C and about twenty frosts occur in this period. With an annual total rainfall of about 546 mm per annum, there is a pronounced summer maximum and droughts often occur in winter. Proportions of rainfall in each

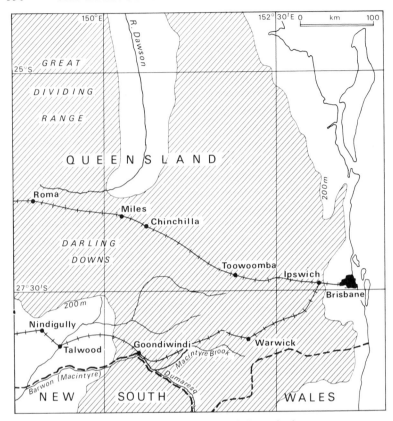

FIG. 65.—*Location of Talwood, Queensland.*

quarter are: spring (September–November), 22.7 per cent; summer (December–February), 38 per cent; autumn (March–May), 22.3 per cent; and winter (June–February), 17 per cent.

The ranch, established in about 1860, was cut out of bush country occupied at that time by nomadic aboriginals. The area is now "loaned crown land". The main improvements since the ranch was first established have been concerned not only with improvement of buildings and facilities, but also with land improvement by timber control and water conservation in order to improve the grazing available.

Livestock and crops

Of the total land area of the holding, 4,304 ha, about 800 ha is cultivated and the remainder is under permanent grass. The pasture can support 1 beast to 3.3 ha, or sheep at the ratio of 1 per 1 ha. The present increased

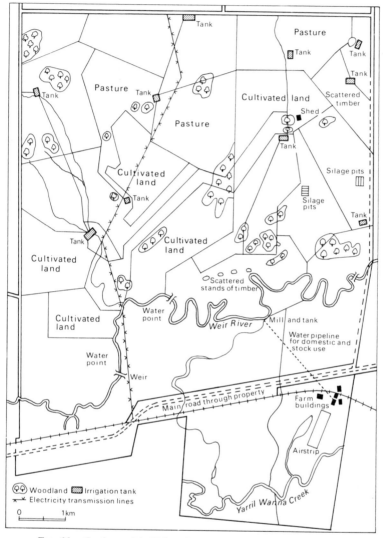

FIG. 66.—*Cattle ranch in Talwood District, Goondiwindi, Queensland.*

importance of wheat, for which the area has been doubled during recent years, is probably only temporary, and has been occasioned by the depressed condition of the world market for cattle at the present time.

There are 237 beef cattle on the holding, predominantly Herefords, including 160 breeding cows and heifers, 70 steers and 7 bulls. There are also 10,112 sheep, consisting of 9,000 breeding ewes, 800 ewe lambs, and 312 rams of the famous Merino breed. In addition 12 horses are employed on the holding.

The bulls are put out with the cows during the summer period (November–March). From about September onwards the cows are calving, and the calves undergo steer selection the following February and weaning in May. The seasons of the sheep management are similar.

Wheat of various varieties is cultivated, and this is practically the only crop. No rotation and no soil dressings are employed. Ploughing takes place in January or February, harrowing in March, and the grain is harvested about May–June. Most of the crop is marketed through the Australian Wheat Board, but some may be utilised for the production of silage, provided the season is favourable and the machinery is available. The yield obtained, approximately 1,100 kg per ha, may be compared with 3,400 kg per ha at the Lac Vert Prairie Farm described in Chapter IX, and with 5,000 kg per ha at Linslade Manor Farm in Bedfordshire, England, described in Chapter XII. Clearly the Talwood

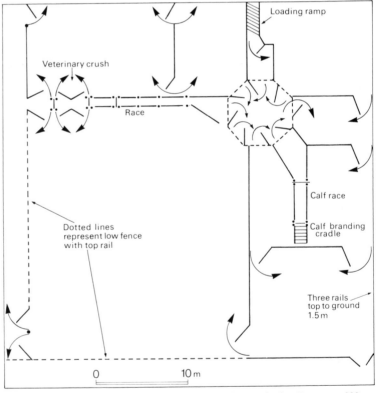

Fig. 67.—*Plan for a typical cattle yard in the Talwood area, for handling up to 200 cows and their calves.*

yield is low even by the standards of extensive wheat production in Canada, but then, as already noted, the cultivation of wheat here is only a temporary expedient, and no fertiliser or rotation is used to improve productivity.

Labour and equipment

There are only two full-time male workers on the entire holding, the owner-manager and one general labourer. Three additional men may be taken on casually to help during two weeks of the year, and contractors are called in for the wheat harvest.

Permanent buildings on the holding include the sixty-year-old farmhouse, and the shearing quarters and stables which are both fifty years old. Most other structures are relatively new, and include a wool shed equipped with shearing plant, an all-purpose shed for grain and machinery, a silo with capacity for 82,000 kg, a machine shed, a hay shed, sheep, cattle and horse yards with drafting races (*see* Fig. 67), and several dams as well as a weir with concrete and earth embankments.

Movable equipment includes a Land Rover, two jeeps, two four-wheel-drive trucks, one caterpillar, two wheeled tractors, two disc ploughs, seed harrows, a four-wheeled trailer 5.5 m long, and a variety of other machines and appliances.

Income

Currently farm income is derived 80 per cent from sheep, 8 per cent from cattle and 12 per cent from wheat. The current poor returns from beef cattle have resulted in a doubling of the area of land used for wheat in recent years, but it is anticipated that this area will again be reduced, and that livestock operations will once more assume a higher proportion of farm activities, when market conditions improve. This demonstrates how the extensive systems are finding that diversification of production provides a hedge against economic fluctuations.

Chapter IX

Commercial Cereal Cultivation

Importance of Cereals

THE word "cereal" has been derived from "Ceres", the name given to
the Roman goddess of corn, harvest and plenty. Cereals are cultivated
grasses and their exploitation dates from Neolithic times, when they
were among the first crops grown by the earliest of cultivators. Even
today cereal crops are the "Staff of life" for many peoples throughout
the world, and indeed, 4 per cent of the world's cultivable area is
devoted to the six leading grains plus potatoes. The world's cereals include wheat, rye, oats, barley, rice, maize ("corn" in the U.S.A.) and
millet.

Of these, rice, maize and millet, which are indigenous to Monsoon
Asia, the Americas and Africa respectively, are grown to provide essential carbohydrates for subsistence peoples. Although produced so extensively, they enter comparatively little into world trade, except in so
far as they are used for commercial livestock fodder, in the case of
maize, or where there is a traffic in rice between deficit producers, such
as India, and surplus producers in the same area, such as Burma and
Thailand. Even then, only 5 per cent of total world rice production
is involved in such international trade.

Of the remaining cereals, barley, although it has several commercial
uses—for fodder, for malting for beer, or in biscuit manufacture—is
of little significance in world trade. The same applies to oats, which
are employed as animal fodder, and also for making porridge, biscuits
and oatcakes, and are grown as a subsistence crop in countries with
a cool, damp climate.

There remain, therefore, only two cereals, rye and wheat, which differ
from all the others in possessing sufficient "gluten" (coagulant) to make
good bread. Rye, however, produces an unpalatable black bread,
whereas wheat bread has an acknowledged universal appeal. Thus,
while rye cultivation is almost wholly confined to the U.S.S.R., Eastern
Europe and some regions of the Americas, and is in the main a subsistence crop apart from such subsidiary uses as for production of whisky
or vodka or for animal fodder, wheat is the supreme commercial cereal.
It is cultivated extensively throughout the Americas, Europe, the
U.S.S.R., Australia and even in parts of Moonsoon Asia. Wheat of
one kind or another is grown under a variety of climatic and soil conditions, and the resultant grains vary in essential properties and uses. The

main distinctions lie between "winter wheat" which is sown in autumn, germinates the following spring and comes to maturity the subsequent autumn; and "spring wheat", which is sown in spring and harvested the same year. The latter is grown under continental climates too severe for winter wheat to survive, such as those of Eastern Europe, the U.S.S.R. and the Americas, and here the resultant flour is "strong", that is, its high gluten content makes it ideal for modern mechanised break-making, or in Italy, where it is grown in the hot, dry summers, for use in macaroni and pasta products.

Winter wheat, grown in countries which, like Britain, have a less severe winter, gives a "soft" flour, which because of its lower gluten content is more suitable for household baking. Production of winter wheat is mentioned in Chapter XII, dealing with intensive rotational mixed farming on the Chelmscote, Broadoak and Linslade Manor Farms.

Canada's Prairie Provinces

In this chapter we are concerned with the production of spring wheat as it is carried on in Canada, which enjoys the distinction of being the world's largest exporter of wheat. The location of Canada's Prairie Provinces is illustrated in Fig. 68.

Ideal Physical conditions for wheat cultivation

These include a fertile, heavy loam soil, a cool, moist season for germination, and a moist, warm growing season culminating in a hot, dry ripening and harvesting season with temperatures of not less than 15°C. There should be a growing season of at least 100 frost-free days, although some strains of quick-maturing wheat have been developed which are capable of maturation in about 60 days and which are suitable for the continuous sunlight of the brief tundra summer as it occurs further north in Canada and the U.S.S.R. Wheat cultivation in the mid-latitudes is largely confined to regions with an annual rainfall of between 380 and 760 mm, since in wetter conditions not only is wheat subject to disease infestations, but also the problem of satisfactory ripening and harvesting is increased.

Advantages offered by Canada's Prairie Provinces

1. *The open, almost flat plains*, gently declining eastwards towards the Great Lakes in a series of steps, are ideal for the application of mechanised techniques of farming to which wheat lends itself, and which, in a country with only a limited labour force and small home consumer market, make possible extensive commercial production of wheat for the supply of world markets.

2. *The severe continental climate* not only has helped in the creation of an ideal soil for wheat cultivation, but also provides suitable weather

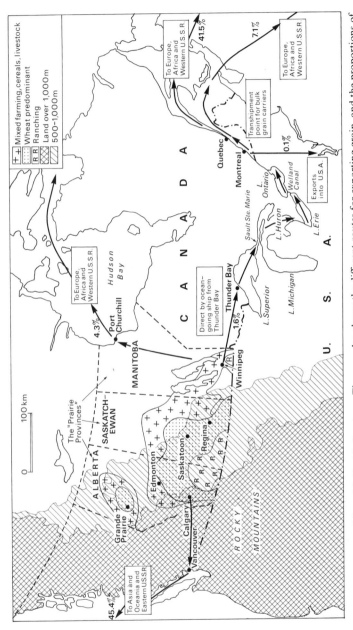

FIG. 68.—*Location of Canada's Prairie Provinces.* The map also shows the different routes for exporting grain, and the proportions of total grain exports they carry.

conditions for wheat cultivation (*see* above). Average climatic statistics for 1966–76 for centres in each of the three Prairie Provinces are given below. These show increased severity towards the west (see in particular the number of frost-free days).

(*a*) *Winnipeg, Manitoba:*

Latitude	Longitude	Altitude	Frost-free days
49° 53′ N.	97° 10′ W.	240 m	118

Mean monthly temperatures (°C):

Jan.	Feb.	Mar.	Apr.	May	June	July	Aug.	Sept.	Oct.	Nov.	Dec.
−18.3	−15.7	−8.1	3.3	10.6	16.5	19.7	18.7	12.6	6.6	−4.4	−13.7

Annual temperature range: 38°C.

Total monthly rainfall (mm):

Jan.	Feb.	Mar.	Apr.	May	June	July	Aug.	Sept.	Oct.	Nov.	Dec.
24	19	26	37	57	80	80	74	53	35	27	23

Total annual rainfall: 535 mm.

(*b*) *Regina, Saskatchewan:*

Latitude	Longitude	Altitude	Frost-free days
50° 30′ N.	104° 38′ W.	574 m	107

Mean monthly temperatures (°C)

Jan.	Feb.	Mar.	Apr.	May	June	July	Aug.	Sept.	Oct.	Nov.	Dec.
−17.3	−14.3	−8.3	3.3	10.6	15.3	18.9	17.9	11.6	5.3	−5.2	−12.9

Annual temperature range: 36.2°C.

Total monthly rainfall (mm):

Jan.	Feb.	Mar.	Apr.	May	June	July	Aug.	Sept.	Oct.	Nov.	Dec.
18	17	18	23	41	83	58	50	36	19	18	16

Total annual rainfall: 397 mm.

(*c*) *Edmonton, Alberta:*

Latitude	Longitude	Altitude	Frost-free days
53° 34′ N.	113° 25′ W.	853 m	102

Mean monthly temperatures (°C):

Jan.	Feb.	Mar.	Apr.	May	June	July	Aug.	Sept.	Oct.	Nov.	Dec.
−16.3	−12.1	−7.3	2.9	9.7	13.3	16.1	14.4	9.7	4.2	−5.6	−12.3

Annual temperature range: 32.4°C.

Total monthly rainfall (mm):

Jan.	Feb.	Mar.	Apr.	May	June	July	Aug.	Sept.	Oct.	Nov.	Dec.
23	20	17	22	35	77	99	66	44	19	19	19

Total annual rainfall: 460 mm.

The significance of the above statistics to wheat cultivation is as follows.

April: the thaw permits May planting and germination with temperature above 0°C.

May–July: rainfall and temperature reach a peak, which is suitable for growth.

August–September: rainfall is declining, while temperatures are still high for ripening and harvest.

3. The deep czernozem ("black earth") soil is rich in humus, nitrogen, phosphorus and potassium. Such a soil is a product of the hot, wet, early summers which stimulate profuse growth, followed by long, severe winter conditions—up to five months with temperatures below freezing—which inhibit growth. The effect over a long period is to bring about a steady accumulation of organic plant nutrients in the top soil horizons. These nutrients are readily available, yet in excess of what may be easily absorbed by the summer growth of any subsequent season. Rich stocks of humus built up in this way have given the "black earths" their characteristic colour.

4. The availability of good routeways for exportation made possible the opening up of the region to commercial agriculture from the 1860s onwards. Particularly important is the route through the Great Lakes, which is, however, blocked with ice for several months, together with the associated railway routes to Montreal. This route was much improved by the St. Lawrence Seaway Scheme, completed in 1959, which enabled wheat from the Prairie Provinces to be loaded directly on to ocean-going ships at Thunder Bay on Lake Superior instead of undergoing subsequent transhipment at Montreal. Other routes have been developed, including those via Hudson Bay and the Pacific Coast. Exports via the latter have increased rapidly, until it now carries the largest proportion carried on any route. The different routes and the percentages of exports they carry are illustrated in Fig. 68.

5. The availability of a large Canadian wheat surplus and a corresponding demand for Canadian wheat overseas are important prerequisites without which Canada would cease to be a major world supplier of wheat. Several European countries cultivate wheat more intensively, and countries such as the U.S.S.R. actually produce far more, but their domestic demands so outstrip their capacity to supply that they must depend upon imports from surplus suppliers such as Canada (*see* Table 5 below and Fig. 112 in Chapter XVIII).

TABLE 5

The chief world producers of wheat, and their status as importers or exporters

Wheat-producing country	Percentage of world wheat production (average, 1966–76)	Percentage of world population	Surplus (+)/ deficit production
1. U.S.S.R.	26.0	6.54	Deficit/importer
2. U.S.A.	13.3	5.53	+Surplus/exporter
3. China	9.0	20.00	Deficit/importer
4. India	6.3	14.20	Deficit/importer
5. Canada	4.7	0.55	+Surplus/exporter
6. France	4.6	1.40	+Surplus/exporter
7. Australia	3.0	0.32	+Surplus/exporter
8. Italy	2.8	1.50	Deficit/importer
9. Turkey	2.8	0.90	Deficit/importer
10. Pakistan	2.0	3.00	Deficit/importer
11. Argentina	2.0	0.64	+Surplus/exporter
12. West Germany	1.9	1.66	Deficit/importer
13. Poland	1.5	0.90	Deficit/importer
14. Others	20.1	42.86	
	100.0	100.0	

Prairie Farm at Lac Vert in North-east Saskatchewan

This farm was established by the father of the present owner, Mr. Lloyd Loyns, in 1903. It is a holding made up of fifteen "quarters", or recti-linear squares, each of 64.75 ha. The total holding of 971 ha is situated at 52° 30′ N. and 104° 30′ W. (*see* Fig. 69).

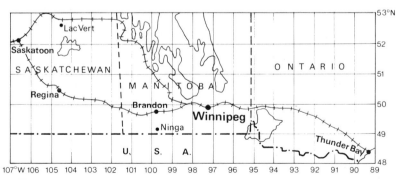

FIG. 69.—*Location of Lac Vert and Ninga Prairie Farms in relation to principal towns and landmarks of their regions.*

Ten quarters are owned, while the other five quarters are rented on a kind of "share-crop" basis, where the landlord receives one-third of the crop less one-third of the costs in lieu of rent. This kind of agreement is increasingly giving way to rents on a "straight-cash" basis.

The farm is situated nearly 200 km further north than Regina, and so the climate is marginally more severe (*see* above for information relating to climate of Regina). The soil is a black silty loam similar to the black Melfort loam which occurs further north. The soil and climate being perfectly adequate, the effective restraint upon farm productivity is market availability. The natural vegetation of these parts, situated in the Parkland region, was originally deciduous forest.

Until about 1968 the farm operation was more diversified than now, with a feed lot for cattle, but fluctuations in the market prices of cattle persuaded the owners to enter the more specialised and labour-intensive area of seed production.

Crop rotation

To avoid overproduction the land is not cropped continuously, although such continuous cropping could well be sustained. At the time of writing, for example, there are still stocks of barley harvested three years previously! To control production, therefore, the following three-year rotation is maintained.

Year 1. Rape-seed or flax-seed is grown to be crushed for oil and meal (the flax is not used for fibre production). The yield of rape is about 2,000 kg per ha, and for flax about 1,300 kg per ha.

Year 2. Barley or wheat is grown, the barley being used for malting or for cattle fodder, the wheat for human consumption, either for the domestic market or for export. Yields for barley are between 4,000 and 6,700 kg per ha, and for wheat 3,400 kg per ha.

> NOTE: These yields for barley and wheat may be compared with those for Linslade Manor Farm, described in Chapter XII, which for spring barley were 4,000 kg per ha, and for winter wheat were 5,000 kg per ha. All these yields represent considerable advances on average yields for the countries as a whole for 1959, as extracted from the *United Nations Year-book for Food and Agricultural Production for 1960*", viz. U.K. wheat yields 3,000 kg per ha, and Canadian yields 1,200 kg per ha.

Year 3. There is a summer fallow and no crop. Weed problems exist, particularly with Canadian Thistle and wild oats.

Marketing

As indicated above the effective restraint upon the farm's productive potential is the limited market demand for the wheat. The following are the two main methods of marketing.

1. The open market for rape, flax, wheat and barley has the advantages that the price fluctuates according to the market and that, there being no quota restrictions, a prudent farmer who sells at the right time can market with a substantial profit. On the other hand, during a period of glut the farmer may find the price severely depressed. In addition, the transport system may become "plugged" with a grain not in demand at that time, rendering it unable to cope with outgoing shipments of grains which are required at that time.

2. The Canadian Wheat Board (C.W.B.) provides facilities for orderly marketing, whereby there are equitable market opportunities for all, and average prices are assured over a given annual period. The grain is collected from producers as required to meet demand, by a quota system. However, the quota system does give an undue advantage to the more southerly prairie-land farmers who operate a two-year rotation, and many farmers prefer to sell at their own discretion when market prices are favourable. Another problem with the quota system is that fulfilment of the quota may involve transporting grain in bad weather.

Management of the farm

The seasonal pattern of farm management is roughly as follows.
    January. The cropping schedule is planned for the coming year. Seed, fertilisers and sprays (pesticides) are ordered.
    February/March. The seeds are cleaned.
    April. Equipment is prepared for the spring work ahead.
    May/June. Seeding and spraying are undertaken.
    July. Painting and building of storage facilities are carried out. Grain is despatched before the end of the crop year, which expires on 31st July.
    August. This is the beginning of the harvest season. All the crops have to be swathed,that is, cut down and allowed to lie in the field for five days or more in order to dry out. Cutting the stubble to a height of 200–250 mm helps to keep the swath above ground so that air can circulate through it and expedite drying. When the farmer is satisfied that the grain is right, it is ready for the combine harvester.
    September. Combine-harvesting of all the crops takes place.
    October. The grain is dried.
    November. Drying of the grain is completed, and the grain is

despatched. Grain can of course be despatched at any time in response to quota demands from the C.W.B.

December. This is the season when the book work, e.g. the tax accounts, is carried out, and when the purchase of new machinery is undertaken. After this a couple of weeks' well-earned holiday is taken over the Christmas season before the new year comes round with the need to plan once more for the seasons ahead.

There are three partners who share the operation of this holding. Mr. Lloyd Loyns, at 53 years of age, is the primary owner, manager and senior partner. He is a graduate in vocational agriculture of the University of Saskatchewan at Saskatoon, and his responsibilities extend throughout the year. He is supported in this by his son Warren, aged 25 years, who is likewise a graduate in vocational agriculture, and additionally holds the B.Ed. of the Saskatchewan University. His duties as a partner in the business involve a share of management and labour between April and December. Ron Loyns, cousin of Warren and aged 23 years, and a graduate of the same university, works on the same basis as Warren.

In addition, there is Rosco Long, aged 57 and an experienced farm-worker over the past forty years, whose main sphere of service is as a tractor driver, and who works on the holding from April to November. Lloyd Loyns's wife also shares in the operation in a supporting role—in the words of her son, she "works the longest hours and gets paid the least!"

Between them they carry out practically all the work of the holding, with the exception of the task of seed-cleaning which is currently put out to contract with a neighbour, although there are plans to take up this business as well.

The farm remains what it always has been, a privately owned family farm. Certain legal conditions in Canada's taxation laws have deterred plans to "go public" and become a corporation. This is by no means a general decision, as many farmers have gone into incorporation because of the immediate savings of a reduced tax-load.

Issues and problems

External political and economic influences, which impinge directly upon the successful operation of his farm-holding, have led to Mr. Lloyd Loyn's active involvement, not only in the political arena, but also in service with the Saskatchewan Wheat Pool, the University Commission, and an agricultural research committee, as well as the local "Save the Rail" Committee.

Some live issues which are of concern to Mr. Loyns, as to all farmers in western Canada, are the consequences of the severe inflation which, by imposing undue difficulties upon all but the largest farmers seeking to buy land, are having long-term effects upon land-ownership and farm size. Farmers are also fighting for the retention of a government subsidy on rail-hauled grain—the so-called "Crow's Nest Rates"—without which farmers in the more remote regions would be seriously disadvantaged. There is also resentment over the fact that quota restrictions prevent farmers from achieving their full market potential, and a controversy has grown up between the exponents of "orderly marketing" and "open marketing". Government policies on railway cut-backs— i.e. the proposals of the Hall Commission on Rail-line Abandonment— are also seen as a threat to some regions, and this raises the question of state ownership versus private control of the railways. There is also concern about market competition between private, foreign-owned companies and farmer-owned co-operatives such as the Saskatchewan Wheat Pool. Finally, devaluation of the Canadian dollar threatens to raise the agricultural costs of production and so reduce the competitiveness of Canadian wheat.

Despite such apprehensions, Mr. Warren Loyns affirms his own faith in Canadian agriculture:

> With changes occurring in every field the farmer has to be well informed, knowledgeable and flexible in order to make the correct decisions at the correct time. With risk and uncertainty a way of life, the farmer has to have a certain type of personality— willing to endure the boom-bust cycles of agriculture. We have endured both and we still feel that the rural life-style is hard to beat!

Mr. Lloyd Loyns has been described by Mr. B. W. Gunn, Corporate Secretary of C.S.P. Foods, as "typical and indeed exemplary of producers across the prairie region" for "his diversified operation, size of operation, and cropping management practices".

Prairie Farm near Ninga in Manitoba

Mr. Stan Hicks operates his cereal-producing holding just west of Ninga, at about lat. 49° N. and long. 100° W. (*see* Fig. 69). The farm occupies a much smaller area than that of Mr. Loyns, consisting of slightly less than three "quarters" of land in two sections, with a total area of just over 190 ha (*see* Fig. 70). This is rather smaller than average for farms in this region. Ninga itself is a small township with a population of about 100 situated about 13 km east of Boissevain and 23 km west of Killarney on the Canadian Pacific Railway line. It has two grain companies, Manitoba Pool Elevators and United Grain Growers. It also has a post office but no store or other facilities.

The climate may best be illustrated by reference to the following statistics for Brandon, which lies just slightly further north.

Brandon, Manitoba:

Latitude	Longitude	Frost-free days
49° 50′ N.	99° 57′ W.	106

Mean monthly temperatures (°C):

Jan.	Feb.	Mar.	Apr.	May	June	July	Aug.	Sept.	Oct.	Nov.	Dec.
−18.6	−15.3	−8.7	2.6	9.9	15.7	18.9	17.7	11.7	5.6	−5.3	−13.8

Annual temperature range: 37.5°C.

Total monthly rainfall (mm):

Jan.	Feb.	Mar.	Apr.	May	June	July	Aug.	Sept.	Oct.	Nov.	Dec.
20	20	24	30	49	76	68	59	39	20	20	22

Total annual rainfall 447 mm.

The region has a rich czernozem or "black earth" soil with high humus and mineral content, based upon a glacial deposit of Waskada

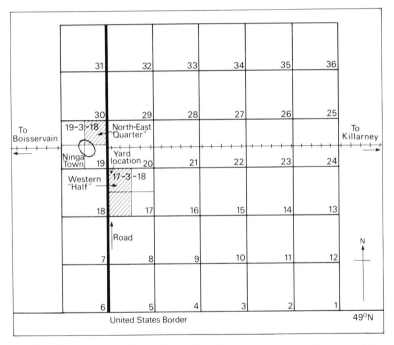

Fig. 70.—*The Ninga Prairie Farm*. The entire grid shown comprises a 6 × 6 mile (9.6 × 9.6 km) block belonging to Range 3 (ranges begin at the U.S. border and extend east–west for 6 miles) and Township 18 (townships extend north–south for 6 miles). Each 6 × 6 mile block is divided into 36 mile-square (2.59 km² or 259 ha) sections. In turn the sections are divided into "quarters", and prairie farms may include non-contiguous quarters in several different sections, e.g. the Ninga Prairie Farm occupies the "North-east Quarter" of Section 19, comprising just under 61 ha because 4 ha are taken up in the town of Ninga, and the "Western Half" of Section 17, comprising 129.5 ha.

Clay, overlying ancient igneous and metamorphic rocks belonging to the Laurentian Shield.

The land here first came into cultivation in the latter half of the nineteenth century when the railways opened up the prairie lands, by giving them access to British urban industrial markets via the Great Lakes.

Crop management

Mr. Hicks grows wheat, barley, oats, rape-seed, flax-seed and maize (corn). Until April snow-cover is heavy and temperatures are still below freezing. After the thaw the snow melts and the soil is saturated, so that the farmer is obliged to wait for the soil to dry out before ploughing can commence.

Mr. Hicks ploughs with an implement called a deep-tiller and refers to the process as tillage. Normally in the spring he deep-tills once, sod-weeds once, seeds the land with the seed-drill and harrows it. The drill also puts on phosphates and the harrow has a tank on it for dressing the soil with a nitrogenous fertiliser if and as required. The process varies according to weed growth, moisture conditions, etc., and indeed every farmer has his own procedure. There is no fixed order for sowing the crops, but there are certain dates by which certain crops are supposed to be sown. Thus the corn-crop has to be in by 20th May. Normally the fieldwork is in full swing by 5th–10th May, although there have been years when spring seeding has been held up until about 25th May. When seeding is as late as this an early-maturing seed is used. A few weeks later, at the end of May or in June, after the crops have broken surface, they are again sprayed with pesticide. A similar rotation is employed as on the Lac Vert holding, except that the land is continuously cropped, and there is no summer fallow on the farm. The precise area of land used for the various crops varies from year to year according to market conditions, weed problems, etc.

The following August or September, once conditions are sufficiently settled and fine and the grain is ripe enough, the crops are swathed, after which they lie for anything from five to fifteen days, depending on the crop and the weather conditions. Combining does not normally start until about noon during harvest time because of the heavy morning dew on the crop, but it may continue until midnight or even longer depending on the weather conditions. The combine picks up the swath, threshes out the grain, and pours it into trucks driven alongside, in one continuous process.

Between completion of the harvest and the Fall, the land is again deep-tilled or ploughed. The Fall comes early in November, and the land remains ice-bound until the following April, so if possible fields of stubble are ploughed before the Fall sets in.

Because of the high cost of machinery, Mr. Hicks operates what is, in effect, a machine pool with his brother, who works an adjacent crop and livestock operation. His own equipment includes a 100 h.p. tractor, an 11 m rod weeder, a combine harvester, a 6 m swather, and one small truck. Normally both seeding and harvesting are joint operations with his brother, who owns another tractor, a 7 m deep-tiller, a 6.5 m seed-drill, a 20.5 m harrow, another combine harvester and a larger 5 tonne truck. This enables Mr. Hicks to manage his holding without resort to outside labour.

Marketing

Mr. Hicks estimates that 40 per cent of his income is derived from wheat, 40 per cent from rape, and most of the remainder from barley. He carries his grain direct to the Country Elevator in his own truck. This Country Elevator is less than 1 km away from the holding, and is one of thousands strung out along the railway lines which thread across the prairie countryside. In the Elevator, the grain is weighed and a sample graded, on the basis of which Mr. Hicks receives a cash ticket from the elevator agent indicating the weight of grain delivered, grade, price, etc., in short, a complete record of the transaction which he can submit to a bank or credit union like a cheque for encashment. A deduction is made at this stage for handling and for the cost of outward transmission to the Terminal Elevator. If, however, it is a glut year and there is too much grain being delivered, farmers may be required to hold back a proportion of it until such time as there is space for it at the shipping points. The grain is taken from the Country to the Terminal Elevators by rail. In the case of Manitoba grain, it is despatched by the eastern route, that is, through Thunder Bay. More westerly provinces despatch their grain through the Pacific ports and various other outlets (*see* Fig. 68).

Mr. Hicks has drawn attention to shortcomings in this system of grain handling, which makes it impractical for the farmers to sell all the grain they can produce. That this complaint is at least in part justified has been borne out by the efforts of the Canadian Grain Council to rationalise their marketing system, and to modernise the facilities for loading, grading and transhipment of grain.

Thunder Bay

The lakeside harbour occupies a site on Lake Superior (*see* Fig. 68), once the site of a seventeenth-century fur transhipment point belonging to the famous Hudson Bay Company. Until 1959 the Great Lakes route was bedevilled by having canals of various sizes, which served to restrict the size of ships and the over-all volume of traffic. Since that date, when it was reopened as a co-ordinated system with canals all capable of taking ocean-going freighters, traffic has expanded accordingly. In 1968 harbour facilities at Fort William and Port Arthur were co-ordinated under the Lakehead Harbour Commission, and in 1970 the towns were renamed Thunder Bay, with a joint population of 110,000. In an eight-month season, Thunder Bay handles some 22 million tonnes of cargo, including 15 million tonnes of grain, 5 million tonnes of iron ore, 500,000 tonnes of general cargo, plus a quantity of newsprint from four modern paper mills. There are also plans to export 3 million tonnes of coal per annum, and the volume of trade is constantly expanding. In addition the port has shipyards and other facilities. However, it does have handling problems associated with the seasonal freeze-up between

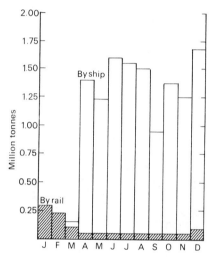

FIG. 71.—*Proportions of grain despatched monthly by rail and water from Thunder Bay in 1973.*

late December and early April, and the equally seasonal character of grain production.

As already noted, Country Elevators impose delivery quotas on the grain farmers at their initial collecting points, but by far the more severe bottlenecks can occur at the Terminal Elevators which currently handle about 15 million tonnes of the grain per annum. Already considerable modernisation and rationalisation of the system have been achieved, and more is proposed. Of an original twenty-six grain elevators the number has been reduced to eighteen, and of these, five are used to handle only speciality grains. It is proposed eventually to switch redundant capacity to loading rail-cars, and it is further recommended that domestic grain should be handled by rail. Remaining elevators will then be modernised to speed up loading procedures and reduce delays. For example, a common loading gallery is proposed which would permit a ship to collect the whole range of grades without moving from one elevator to another, thus reducing loading time. This becomes increasingly essential as 25,000 tonne bulk grain carriers become commonplace. There is a comparable case for rationalising railway loading procedures and rolling stock. Such changes will reduce delays, and a round-the-clock shift system is also recommended to increase over-all capacity.

The basic problems of seasonal operation are, however, inevitable. Between late December and early April the harbour at Thunder Bay and the entire St. Lawrence Seaway are impassable because of the

FIG. 72.—*Modern bulk-carriers on the St. Lawrence Seaway.*

freeze. The small proportion of total grain traffic travelling at this season is by rail. This restraint is not conducive to economic rail operation either, since although a small volume of grain travels from Thunder Bay by rail at all seasons, the bulk of rail shipments occurs during the freeze (*see* Fig. 71).

On average, of annual grain shipments from Thunder Bay over the period 1965–74, totalling 12.2 m tonnes, some 95.5 per cent was despatched by ship, the remainder by rail. Grain travels either by bulk-carriers, which are specialised lake transporters designed to maximise their use of the canal capacity with dimensions of 225.5 m in length by 23 m in width, down to Montreal (*see* Figs. 72 and 73), or by a certain number of ocean-going ships able to make the journey to Thunder Bay economically, load with grain and make a direct return voyage to European or other destinations. The proportion of such ocean-going ships arriving to load with grain is small, but on the increase. As already indicated, a total shut-down of such operations occurs in winter.

FIG. 73.—*Modern bulk-carrier and lock on the St. Lawrence Seaway.*

All the above marketing details might appear irrelevant to a consideration of agricultural production, but they serve to demonstrate how technological achievements have provided the once inhospitable and inaccessible continental interior with access to world grain markets, and yet how other limiting factors still place a restraint upon its potential. In order to correct what might otherwise mislead the student, it must be stressed that, at the national level, Canadian farmers enjoy a substantial degree of security as a result of their Government's efforts to protect them from price fluctuations on the international grain market, and that the ultimate restraint upon grain production is, of course, a limited world demand and not the shortcomings of the Canadian marketing system.

The Canadian Government has been obliged to encourage prairie farmers, particularly those operating in the marginal areas, to grow less wheat by diversifying their production or by switching to other crops altogether. This policy has been undertaken to restrain overproduction, which might otherwise reach a scale at which, if the Government were to continue its policy of price support, even more costly long-term grain storage would be required than at present.

Chapter X

Commercial Plantation Agriculture

Distribution, Development and Characteristics of the System

As already noted, commercial plantation agriculture is essentially a European type of agriculture mainly imposed upon tropical regions (*see* Fig. 20). It was given its initial impulse by the discovery of sea routes to the New World, West Africa and the Far East by Spanish and Portuguese explorers, and subsequently by British and Dutch explorers, in the fifteenth and sixteenth centuries, rapidly followed by the establishment of colonies and trading posts in the regions thus reached. In some cases the existing inhabitants were sufficiently advanced to provide a source of indigenous labour, in which case they were often ruthlessly exploited if not enslaved by the Europeans, and set to work cultivating various tropical crops—mainly foodstuffs—for the European consumer. Where, on the other hand, they were too few in number or technologically too backward or inept to be utilised in this way, they were elbowed aside, assimilated or even exterminated by their conquerors, and more advanced tropical labour was introduced from elsewhere, the classic example being the importation of West African slaves into various parts of the New World over a period of about 300 years. Much later plantation development received further impetus from the Industrial Revolution in the eighteenth and nineteenth centuries, when industrialisation led to an unprecedented rise in demand for industrial raw materials such as cotton, and for tropical food products to feed the growing urban population. As enlightened public opinion in Britain and the U.S.A. began to turn against slavery as an institution, so commercial interests adopted more subtle forms of exploitation such as the importation of indentured Indian workers to work on the plantations of East Africa, Malaya and the East Indies. In the U.S.A. the plantation system collapsed as a result of the abolition of slavery, but some economic exploitation has lingered down to recent times.

Recent years have seen the gradual self-emancipation of tropical peoples, under stimuli such as new nationalist aspirations, the extension of educational opportunities, and the introduction into tropical lands of ideas of organised labour. In newly emerged countries the plantation system has been modified or radically changed, with a growing share of ownership and investment being in the hands of tropical people, who have invaded the managerial and technical grades of the labour force once reserved for Europeans. In some countries, for example in Malaysia, the governments have favoured smallholder production, regarding the prosperity of a self-employed independent agricultural class as providing a more effective bulwark to democracy against threatening communist infiltration than that afforded by a population of wage-earning plantation workers.

General characteristics of the plantation system

1. Monocultural production is practised, that is, one crop only, such as coffee, tea, cocoa, sugar, cotton, rubber or tobacco, is produced.
2. Plant is required for processing, grading, packing and despatch of the product. The plantation may also undertake the provision of basic amenities, such as roads, railways or mains drainage, where these do not already exist in the region.
3. Enormous overheads are involved, not only in developing the necessary amenities described under 2. above, but also in assembling together a large, skilled, often multiracial labour force to supplement the scant labour force available locally, and providing each such ethnic group with housing, social, medical and educational amenities appropriate to its needs.

From the above points, it is clear that plantation production is particularly inelastic (*see* p. 85), whereas the world markets for agricultural products tend to display considerable elasticity. Once established, it is extremely costly, because of the huge outlay and overheads, for a plantation to regulate, let alone suspend, production. It is cheaper even to produce at a loss. Closure means disbanding entire communities of employees, and leaving expensive plant and amenities in wasteful inactivity. It is also difficult and expensive for a plantation either to diversify production or to switch entirely to production of another crop. Physical conditions are not always appropriate or the amenities already established are too specialised for ready adaptation. This must not be exaggerated, however, since plantations in Ceylon (now Sri Lanka) switched entirely from coffee to tea production in 1869, while in the last few years Malaysian plantations have successfully abandoned monoculture for mixed cropping. Recent changes in Malaysian plantation production are considered subsequently in this chapter.

Because plantation agriculture is geared to supplying European markets and because it is a commercial system, it tends to be "Western" and "alien" to the character of the culture and people and to the regions amongst which it is found. In early times many remarkable mistakes were made by plantation managements who did not fully understand the environment or the people they were dealing with. For example, the common mid-latitude practices of weeding the growing crops and breaking up the soil were found to cause soil erosion when they were adopted in humid tropical regions. Over the years plantation enterprises have learned to evolve new agricultural techniques, to cope with tropical diseases in plants and man, and to deal with social problems among their labour force which have required the rethinking of basically European ideas.

Malaysian Rubber Production

Physical conditions in Malaysia

Since the region under consideration is so near the equator, for example the plantation described subsequently, Kuala Jelei, is situated at about lat. 2° 45′ N. and long. 102° 25′ E., the climate is most readily compared to that of Singapore. There the heat, rainfall and humidity are all practically constant, with a mean annual temperature of 27°C, and a mean

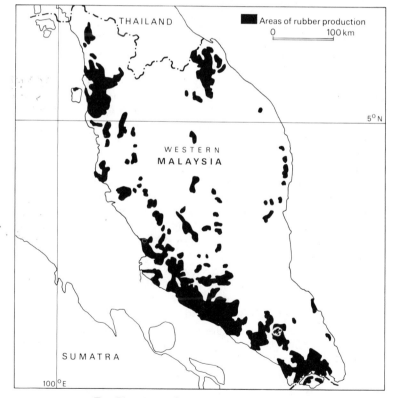

FIG. 74.—*Areas of Malaysian rubber production.*

annual temperature range of 1°C. Rainfall displays a slight November–February maximum, with 760 mm falling in this season out of an annual total rainfall of 2,410 mm. These conditions accord well with the needs of the rubber plant for heavy rainfall of at least 1,500 mm per annum and temperatures of not less than 20°C, preferably averaging 27°C.

Rubber will grow on a wide variety of soils including coastal clays, alluvium and quartzite or shale soils, and is even tolerant of laterites; indeed rubber is only wholly contra-indicated on peaty or badly drained soils. Preferably, rubber requires an extensive lowland with deep, rich, well-drained soils, and it is significant that production in Malaysia is largely restricted to the west-coast area of the country where all these conditions are met (*see* Fig. 74).

The development of the Malaysian rubber industry

The most common source of latex rubber is the tropical plant *Hevea brasiliensis*. As its name implies, it is a native of Brazil, and in the nineteenth century the then small world market for natural rubber was practically a Brazilian producers' monopoly. At that time it was gathered from wild trees in the Amazon Basin. Today, however, Malaysia alone produces more than 43 per cent of the world's supply. In about 1876 some Britons smuggled seedlings of rubber out of Brazil, and after propagation at Kew Gardens, London, plants were sent first to Ceylon (now Sri Lanka), and subsequently to British Malaya.* Here they were of only limited interest at first, but were cultivated as a plantation crop from the 1890s onwards (between 1876 and 1890 coffee was considered of much greater importance). It was about this time that several factors stimulated the growth of a world market for rubber which Malaya, as a plantation producer, was better equipped to exploit than were the wild-rubber barons of Brazil. Notably it was the development of the pneumatic tyre, the advent of the motor car, and the increased use of electric wiring and cables which require effective insulation that contributed to the growth of the Malayan rubber industry. During the Second World War, however, both the U.S.A. and Germany developed synthetic rubber industries, and by the end of the war could produce a quantity equal to the world's output of the natural product. Today, synthetic rubber production far outstrips that of natural rubber, but in absolute terms, Malaysian production has still increased to meet a growing world demand. Natural rubber still has properties such as greater elasticity and resistance to tearing which recommend it for tyre manufacture in particular. Tables 6 and 7 illustrate the current position of natural rubber and Malaysian production in particular in the world market.

* British Malaya achieved independence in August 1957. In September 1963, the new independent federation of Malaysia was formed from Malaya, Singapore (which later left), Sarawak and Sabah (North Borneo).

TABLE 6

The place of Malaysia in world production of natural rubber

	World production (tonnes)	Malaysian production (tonnes)	Malaysian as percentage of world production
1967	2,522,500	990,446	39.3
1972	3,120,000	1,304,147	41.4
1977	3,600,000	1,613,193	44.8

TABLE 7

Relative importance of natural and synthetic rubber on the world market

	Total production (tonnes)	Natural as percentage of total	Synthetic as percentage of total
1967	6,872,500	36.7	63.3
1972	9,885,000	31.5	68.5
1977	11,982,500	30.0	70.0

The main producer of synthetic rubber in 1967 was the U.S.A. with 44.7 per cent of world production. Since then world production has risen by 30 per cent, but the U.S. share has declined to 28.8 per cent. Other producers include Japan, France, West Germany, the U.K., Canada and the U.S.S.R.

There has, however, been another domestic upheaval in the Malaysian rubber industry itself of which the above statistics show no indication. Over the same period smallholder production of rubber has overtaken plantation production both in area under rubber and in output. As early as 1958 the smallholders between them had a larger area of land under rubber than the plantation producers, but because their trees were on average older and their methods of production less efficient, it has taken until 1973 for them to overtake plantation production, and even their share of total output, 54 per cent, has to be seen against their control of 64 per cent of the land under rubber. In 1973 their unit yields of latex per land area were only a little more than half that obtained on plantations, and although they have subsequently been able with government help to replant a much larger area than before with the new high-yield rubber trees, they are still behind the plantation producers, who continue to lead the way in high-yield cropping.

The Dunlop Estates in Malaysia

In recent years the plantation interests have had to adapt themselves to the general decline of European influence and the emancipation of indigenous tropical populations. The achievement of Malayan Independence in 1957 made their interests insecure, and this, on top of the rising costs and huge overheads they had to face, made future prospects uncertain, but by dint of their adaptability to political and technological change, they have been able to survive.

Dunlop Estates Berhad was incorporated as a public company in 1967 and acquired the undertaking and assets of Dunlop Malayan Estates which had previously handled the production side of Dunlop Ltd., who had had extensive interests in Malayan rubber since 1910. Dunlop Estates Berhad has 4,000 local investors, including individuals, corporations and trust funds.

The company owns ten estates in West Malaysia, which include five in northern Johore, four in Negri Sembilan and one in Malacca, with a total area of 23,806 ha under cultivation as follows:

		hectares	*hectares*
Rubber:	⎰mature	12,188	
	⎱immature	40	
	TOTAL IN RUBBER		12,228
Oil palm (since 1965):	⎰in bearing	7,727	
	⎱immature/clearings	1,015	8,742
Cocoa (since 1969)):	⎰bearing	919	
	⎱immature/clearings	726	1,645
Coffee:	immature	—	2
	TOTAL PLANTED/CLEARED		22,617
	Roads/sites, etc.		1,189
	TOTAL AREA OF PLANTATIONS		23,806

The diversification into oil palm, cocoa and coffee shown in the above table has been undertaken to reduce the dependence of the company upon rubber. However, in this account attention will be confined primarily to developments in the production of rubber. Figure 75 shows the location of the ten Dunlop estates, including one relatively small rubber and oil-palm plantation at Kuala Jelei which is selected for more detailed study.

In all, Dunlop Estates Berhad, Malacca, operates various plants for the processing of rubber, cocoa and palm oil. As regards rubber processing, these include three large factories which produce concentrated latex, a Hevea-crumb factory producing Dunlocrumb S.M.R. 10, two factories for production of S.M.R. 5C.V. by the comminuting process, a factory for skim crumb, and a Research Centre at Betang Malaka responsible for the "quality control" of the company's products. The Dunlop Research Centre maintains close liaison with the Rubber Research Institute of Malaysia and other Dunlop Research Centres

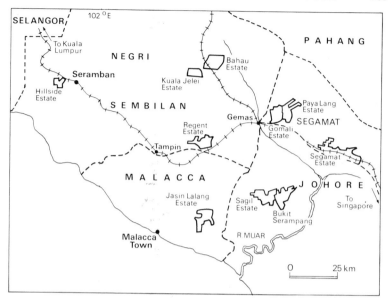

FIG. 75.—*The Dunlop Estates in Malaysia.*

elsewhere in the world. The company also operates storage facilities at estates and railheads, and port bulking facilities, for the handling, accumulation and distribution of cargoes of rubber as well as palm oil.

Since 1948 the company has replanted 12,500 ha of high-yield rubber, this being essential to enable plantations to compete with smallholders, who have economic advantages in other ways, as indicated in the conclusion to this chapter. Yields of rubber in 1977 were 1,770 kg per ha. The effects of diversification included a slight decrease in over-all rubber production from 21,663 tonnes in 1976 to 21,472 tonnes in 1977. This decrease was negligible in view of a reduced area of 506 ha. Concentrated latex accounted for 63 per cent of the output of 1977.

Dunlop pioneered bulk exportation of concentrated latex back in the 1930s and indeed accounted for nearly 10 per cent of the total concentrated latex exported from Malaysia in 1976. The products of Dunlop Estates Berhad, Malacca, are exported to both the E.E.C. and Comecon countries of Europe, and to Russia, Japan, Korea and Australasia.

Other smallholders and estate producers in the vicinity of Dunlop Estates often market their product through the company, and in 1977 rubber from these sources included 7.7 million dry kg of field latex for concentration. This makes for far more centralised, efficient processing and marketing, and enables the smallholder to obtain a fair market price for his product without laying out capital on expensive processing plant.

The labour force on the Dunlop Estates exceeds 6,000 and there are in all 13,000 workers and dependants accommodated by the company. Facilities provided by the company include those for sport, scholarships for primary and secondary education, and various kinds of staff training and improvement schemes.

The company's future is bright in view of its successful crop diversification policies, the development of high-yield rubber and the concentrated latex production. Nevertheless, while plantation production in general may be able to adapt to socio-economic changes, long-term prospects for rubber production are open to question.

For the above information and for the following study the author is indebted for the kind assistance afforded by the Rubber Research Institute of Malaysia and by Dunlop Estates Berhad, respectively.

The Kuala Jelei Estate, Negri Sembilan, Malaysia

The information embodied in this study has been furnished by the estate's manager, Mr. G. M. McManaman.

The Kuala Jelei Estate is one of ten plantations owned by Dunlop Estates Berhad and is situated in the Bahau sub-district in the State of Negri Sembilan, about 73 km east of the State capital of Seramban. The estate contains 682 ha of undulating ground between 30 and 60 m above sea-level, rectangular in shape and situated mainly on the west bank of the River Muar. A plan is given in Fig. 76.

The estate operates independently for the production of rubber and fresh fruit bunches of the oil palm, but the subsequent processing and despatching is carried out at other estates within the company.

The Bahau sub-district was once tropical forest, but this original vegetation has now been felled except for a forest reserve on a steep ridge which runs from north-north-west to south-south-east and rises to an altitude of 361.5 m at the peak of Jeram Padang South. This jungle-covered ridge is situated on the eastern boundary of the estate, and provides the water catchment for part of the estate's water supply. A number of streams drain across the plantation into the River Muar, so that the entire area enjoys good natural drainage and all the land can be utilised.

The predominant soils on the estate are Haplorthox/Oxisols and Palcudults/Ultisols, with a fair amount of alluvial material from the adjacent river. The pH reading of these inland soils is between 4.5 and 5.7 (acidic).

The relief necessitates construction of terraces along the hills and ravines in order to conserve moisture and prevent soil erosion. Soil fer-

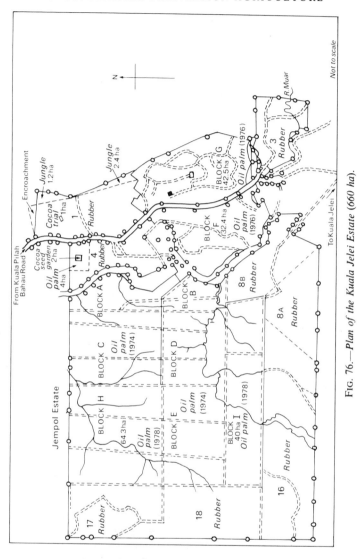

FIG. 76.—*Plan of the Kuala Jelei Estate (660 ha).*

Palm oil: total 380 ha
  Blocks A, B, C, D, E planted 1974 (197 ha)
  Blocks F and G and small 4 ha plot planted 1976 (79 ha)
  Blocks H and I planted 1978 (104 ha)
Cocoa: total 3 ha
  2 ha cocoa seed garden
  1 ha cocoa trial
Rubber: total 277 ha
  Sections 1, 3, 4, 8A, 8B, 16, 17, 18 planted 1949, 1954, 1959, 1971
  Unproductive land, roads, etc: total 22 ha

tility is not a problem since a discerning fertiliser policy is practised, based upon chemical analyses undertaken by the company's research centre. The physical quality of the soil, on the other hand, is important, and can be improved by careful husbandry methods.

Rubber was first planted at Kuala Jelei in 1918, but the planting was not completed until Dunlop bought the estate in 1927. Until 1973 the entire estate was planted in rubber, but since 1974 a large proportion has been replanted with oil palm (*see* Fig. 76) and by 1984 the estate will have only 20 per cent of its land area under rubber, as shown in the following figures:

1970	100%	Rubber	660 ha
1978	42%	Rubber	277 ha
	57.5%	Oil palm	380 ha
	0.5%	Cocoa	3 ha
1984	20%	Rubber	132 ha
	79.7%	Oil palm	526 ha
	0.3%	Cocoa	2 ha

The reason for the decline in land area under rubber here and elsewhere is because palm oil is almost twice as profitable to produce as rubber. Rubber yields M$990–1,100 per ha on average, but oil palm yields M$1,980 per ha.

The type of rubber clone selected for planting is assessed on the basis of the "environmax" concept, i.e. clones are chosen according to their compatibility with the complete environment, taking into account such aspects as soils, topography, exposure to winds, rainfall distribution and vulnerability to leaf and stem diseases, e.g. *Oidium heveae* (mildew) and *Gloeosporium alborubrum* (leaf spot).

Kuala Jelei now has 277 ha of rubber, all of which is in tapping, and the staff have to ensure that tapping commences early in the morning when rubber latex flows most freely (*see* Fig. 77)). It is also important that tapping is done properly with the correct angle of tapping cut, at the best depth, with necessary economy of bark removed, and that clean utensils are used to collect the latex. Failure to tap skilfully can reduce the economic life of the tree. Maximum latex collection may be achieved by tapping some kinds of tree every other day, others by tapping once every third day. Staff are required to report trees which have been struck by pests or disease, fire or lightning damage, or which require treatment of some kind.

Although the life span of *Hevea brasiliensis* is 90–100 years, economic life span is only 25–30 years. Research is currently being directed towards trees being brought into rubber earlier than hitherto and maximising yields, even though this shortens the productive life of the tree. Most of the latex tapped at Kuala Jelei is used to produce concentrated latex, although some lower-grade latex, e.g. cup lump or tree lace, is used for Standard Malaysian Rubber (S.M.R.). The latex concentrate goes by road and rail to Singapore, the S.M.R. by road to Port Klang. While the bulk of latex concentrate is exported to Britain, both these products are despatched to regions all over the world.

If any terrain looks susceptible to soil erosion after clearing, trees are planted along contours and anti-erosion terraces are constructed.

Fig. 77.—*Collecting latex.*

Each soil, after due analysis, is given its distinctive fertiliser schedule by the Dunlop Research Centre, indicating the appropriate type and quantity of fertiliser per tree for that particular section of the plantation.

Apart from replanting, when heavy equipment is employed to fell and stack the trees, the most heavily used piece of equipment is the tractor, which is employed for carting fertiliser, for carrying water for the sprayer, and for sulphur dusting against leaf diseases such as mildew and leaf spot.

The staff at Kuala Jelei, apart from the manager who is British, are all Malaysians employed through contractors, and all full-time (*see* Table 8). Of these, 86 per cent live in free company accommodation, the remaining 14 per cent nearby in their own houses situated in surrounding villages.

TABLE 8

The labour force at Kuala Jelei

Grades of employment	Ethnic origins of the labour force						Occupation totals
	Malay		Chinese		Indian		
	Male	Female	Male	Female	Male	Female	
Executive assistant	1	–	–	–	–	–	1
Clerks	1	–	–	–	2	–	3
Lightset operator	–	–	–	–	1	–	1
Lorry driver	1	–	–	–	–	–	1
Lorry attendant	1	–	–	–	–	–	1
Rubber tappers	22	16	7	17	8	4	74
Oil-palm harvesters	9	4	–	–	1	–	14
General workers	8	11	5	2	2	4	32
TOTALS	43	31	12	19	14	8	127

NOTE:
Malays make up 58 per cent, Chinese 25 per cent and Indians 17 per cent of the labour force. General workers include weeders, sprayers and pollinators. The Kuala Jelei Estate operates one lorry used for transporting field-latex to the Bahau Estate each day for processing in their latex concentration plant.

Conclusion: Future Prospects for the Industry

Since the Malaysian Government sees the smallholder as an independent, self-employed proprietor who is potentially its most effective bulwark against the inroads of communism, it is disposed to favour this type of agriculture in its tax system. This, coupled with the political risks and large overheads confronting the plantation operators, makes long-term prospects poor. In the words of Mr. McManaman of Kuala Jelei, "Unless Malaysia changes its tax system, rubber will eventually become a smallholder's crop." It has been this consideration, as well as the higher profitability of palm oil, which has encouraged diversification of production, adoption of high-yield rubber trees and of techniques for maximising latex extraction, as well as the development of latex concentrates, all calculated to ensure at least the short-term future of the plantation industry in Malaysia.

In the long term, it seems almost inevitable that although the research "spin-off" of plantations, and the marketing facilities they extend to the smallholders, actually are beneficial to them, the smallholders will in time drive the plantations out of business. As the smallholders become more efficient producers and become commercialised instead of carrying out subsistence farming as at present, they and some medium-sized estates not dependent upon foreign capital will become the "norm". Alternatively, rubber producers' co-operatives could develop, or collectives in which the rubber growers operate a large unit under an elected management. This depends as much upon political as upon economic or technological changes in the future.

Chapter XI

Mediterranean Agriculture: Subtropical Fruit and Grain Cultivation

Introduction

THE regions of mediterranean climate all lie between latitudes 30° and 45° N. and S., and are transitional in character between the mid-latitude, cool, temperate regions on the one hand, and the tropical, hot, desert regions on the other. The climate is dominated by the cyclonic conditions of the former in winter, and the tropical, high-pressure wind systems of the latter in summer, giving rise to cool, damp winters and hot, arid summers. Such generalisation, however, conceals considerable variations in what is a transitional system, and which, along with other physical and economic influences, help to produce a wide range of agricultural responses.

The main regions of mediterranean climate include not only those regions which fringe the actual Mediterranean Basin of Southern Europe, but also the south-western cape of South Africa, the extreme south-west and south-east regions of Australia, central Chile, and lower California, U.S.A. The comparatively late European settlement of the latter regions, however, has led to the achievement of commercial fruit and wine production, and of market gardening, far in advance of that reached in the Mediterranean Region of Southern Europe, as described later in this chapter with reference to the Maltese islands. In Chapter III some reference has been made to agriculture in the Great Valley of California, while Chapter XIV looks at agricultural developments in Israel, under the most extreme of Eastern Mediterranean conditions. The statistics in Table 9 are included to demonstrate the basic similarity not only between mediterranean climates in the east and west sectors of the true Mediterranean Basin, but also between mediterranean climatic regions all over the world.

The agriculture of the Mediterranean Basin is still largely underdeveloped, not only because of the environmental problems which have to be overcome, but also as a consequence of the low level of industrial development in the area which renders the agricultural sector the more impoverished, since it is obliged to support a large proportion of the population, yet lacks the benefit of remunerative markets close at hand which are enjoyed by farmers in the industrial countries.

In this chapter studies are included to demonstrate the problems of farming in a difficult environment without the economic backing and

TABLE 9

The climate of mediterranean regions

Location	Altitude (metres)	Mean temperature, (°C)		Rainfall	
		Winter (*Jan. in north, July in south*)	Summer (*July in north, Jan. in south*)	Per annum (*mm*)	Summer quarter (*mm*)
NORTHERN HEMISPHERE					
(Oct.–Mar.: cool, damp; Apr.–Sept.: hot, dry)					
Lisbon 38° 5′ N. (West Med.)	95.4	11	23	729	22
Athens 38° N. (East Med.)	107.0	9	28	391	30
Los Angeles 34° N. (California)	95.1	13	23	381	0
SOUTHERN HEMISPHERE					
(Apr.–Sept.: cool, damp; Oct.–Mar.: hot, dry)					
Cape Town 33° 30′ S. (South Africa)	17.1	12	21	635	51
Adelaide 35° S. (S. Australia)	42.7	11	23	533	61
Valparaiso 33° S. (Chile)	41.1	11	17	508	15

security which would accrue in a more prosperous situation, as typified by farming in the Maltese islands of Malta and Gozo.

The Maltese Environment

The Central Mediterranean islands of Malta and Gozo are situated 93 km due south of Sicily. The largest of the islands, Malta, extends 27 km along its long axis which trends west-north-west, east-south-east, is 14.5 km wide, and covers an area of 246 km². Gozo, about 5 km to the north-west, is 14.5 km from west to east and up to 6.4 km from north to south, with an area of some 67 km². Between them the tiny island of Comino occupies an area of only 2.6 km².

The geology, structure and relief

The islands are composed of Tertiary limestones with subsidiary clays and marls. There are superficial deposits of the Pleistocene Age consisting of cliff breccias, cave and valley loams, sands and gravels.

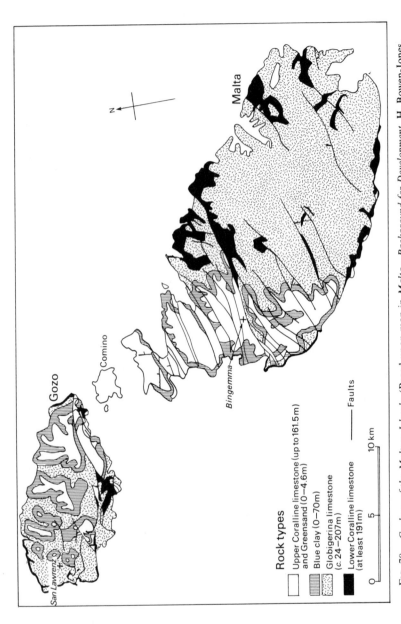

Fig. 78.—*Geology of the Maltese Islands*. (Based upon map in *Malta—Background for Development*, H. Bowen-Jones, J. C. Dewdney and W. B. Fisher.)

The geological structures are exceedingly complex, involving numerous faults which defy simple description. However, the main topographical elements can be summarised as follows (*see also* Fig. 78).

1. There are high Coralline Limestone plateaus.
2. The plateaus are bounded by scarps which display Blue Clay slopes.
3. "Rdum"-type coastlines exist, particularly along the south-west littoral of Malta, where there are regions of specialised agriculture clinging to narrow terraces on cliff platforms which run parallel to the coast.
4. Flat-floored basins have been formed by faulting, down-warping or occasionally erosion, and have made possible the accumulation of younger alluvial deposits.
5. Globigerina Limestone hills and plains give rise to a gently undulating topography.

The climate

The climate may be illustrated by reference to the following statistics for Valletta.

Latitude	Longitude	Altitude
35° 54′ N.	14° 31′ E.	56 m

Mean monthly temperatures (°C):

Jan.	Feb.	Mar.	Apr.	May	June	July	Aug.	Sept.	Oct.	Nov.	Dec.
11.7	11.7	12.8	15	17.8	21.7	25	25	23.3	20.6	16.7	13.3

Annual temperature range: 13.3°C.

Total monthly rainfall (mm):

Jan.	Feb.	Mar.	Apr.	May.	June	July	Aug.	Sept.	Oct.	Nov.	Dec.
88.9	55.9	40.6	22.9	10.2	2.3	1.0	5.1	27.9	78.7	91.4	96.5

Total annual rainfall: 521.4 mm.

These statistics are, however, subject to considerable seasonal and regional variation, particularly as regards rainfall. Thus annual rainfall has been recorded as high as 998 mm and as low as 200 mm. Individual months have shown similar deviance, e.g. October, which averages 79 mm of rain, has actually experienced as much as 586 mm in 1951, and no rain at all in 1859. Such unpredictability of rainfall, as expressed in periodic droughts and floods, yet coupled with uneven distribution seasonally and regionally, high levels of summer evaporation, run-off, and loss of moisture because of the porosity of many underlying rocks, all make irrigation essential to the maintenance of agriculture.

As regards temperature variability, ground frosts may occur on only a few nights of a given year, but plant growth may be checked by low winter temperatures, while high spring and summer temperatures may give rise to severe pressure on supplies of water available for irrigation purposes. Locally, temperatures may deviate from those in Valletta by as much as 2.8°–3.3°C.

A distinctive feature of the Maltese climate is the occurrence of violent storms which may account for as much as 10 per cent of the annual rainfall, and as a result of which daily falls of 50–75 mm may occur. Such rainfalls occur on about 77 days of the year, but particularly in the winter season. It is quite impossible to conserve the rainfall released from these violent storms, which are so localised in their effect and bring damage to terraces, soil erosion, gullying, flooding of lowland, and loss or damage to crops in their train.

A more valuable source of water is afforded by the rain brought to the islands in winter by the predominantly north-westerly winds which prevail on 29 per cent of the days in an average year, the rainfall of which is accentuated by the orographic effect of steep north-east trending coasts and highlands. Valletta is on the drier, eastern flank of Malta.

The rainfall of the island, subject as it is to seasonal and regional irregularity, but also effectively minimised in agricultural value by excessive evaporation, run-off and soil porosity, requires supplementation by irrigation. Techniques employed for irrigation purposes include watering by hand from cisterns cut into the rock on the drier Globigerina lands of southern and western Gozo and central and eastern Malta. On the Upper Coralline plateaus pumps operated by wind, donkey or machine power are employed, although where too many such pumps are installed this has the effect of lowering the water-table. In the Lower Coralline and Globigerina there are pumps which draw upon the sea-level water-tables, and which may on occasions so lower the water-table that saline water is brought up. Finally, along the junction of the Upper Coralline with the Blue Clay there are spring-lines from which water is readily available to the farmer.

It is estimated that more than 9,000 million litres of water per annum is used for the irrigation of agricultural land from these sources.

The soils

Soils are the product of their parent materials, and of the impact upon these of climate over time, modified by topographical and biological influences. In Malta, according to one authority, the soil types are only of a limited variety because they are derived from similar parent materials—principally those derived from limestone rocks. Protracted summer drought restricts vegetal growth, and thus the arid conditions preclude accumulation of humus. Consequently only in a few exceptionally wet locations such as occur in deep-set valleys do humus horizons develop, and this further restricts soil diversification. Finally, soil development is slow in limestone country under such arid conditions, and so most soils are immature.

The topography of Malta has influenced soils mainly by causing gully erosion and a redistribution of calcareous materials as a result of the violent rainstorms of early winter. Biological factors are of negligible effect in this arid environment, but the influence of man stretching back into prehistory has been profound. In some cases soils have been transformed as a result of the accumulation of added limestone in the form of rock flour and rubble derived from quarrying and terrace construction. Man's influence has been the more profound because natural processes have been so slow.

The soils include the following. (*see also* Fig. 79).

1. Carbonate Raw soils consist of sandy to clay soils containing a high proportion of calcium carbonate, and calcite sand, directly based upon the underlying parent rocks, either limestone or the adjacent Blue Clay. There is only 1–2 per cent of raw humus near to the surface of these soils, which have a very high pH value, about 8.5, indicating their highly alkaline nature. The more sandy of these soils are dry, easily worked, and are sufficiently fertile for vines and fruit trees to be cultivated. The clay loams and clays, however, are water-retentive and so inclined to go hard and dry in late summer. They are utilised for a wide variety of crops, including sulla, vines, tomatoes, cereals and potatoes. Most of these soils either occur on steep slopes or have been terraced.

2. Xerorendzinas consist of very fine, grey, dry, ash soils, containing between 1.5 and 6 per cent of humus and 50–80 per cent of calcium carbonate. They display a pH of 8.35–8.55, very slightly less alkaline than the Carbonate Raw soils. Although such Xerorendzinas appear to be so chemically deficient, they are extensively cropped, with cereals, vines, roots and market-garden produce on the more sandy soils, and cereals and potatoes on the clay soils.

3. The Terra soils, in contrast to those mentioned above, are "very mature, extensively weathered" and are coloured from yellow through brown to red with ferric hydroxide, colour variations being related to water content. These soils are richer in humus, averaging 4.5 per cent and are relatively less alkaline, displaying a pH of about 8.0. In texture they range from loose, dry, friable loams to heavy viscous clays. Where Terra soils occur on rocky limestone surfaces, they are dry and practically unworkable, and so are usually given over to poor rough grazing for sheep and goats. On arid hill slopes they may be terraced to provide poor crops of cereals and potatoes, although the process of terracing may add so much raw calcium carbonate in the form of rock flour, that many such areas of man-made soil complex have had to be abandoned subsequently as uneconomic to cultivate. Where, in contrast, the soils are partially the result of alluvial deposition in well-watered

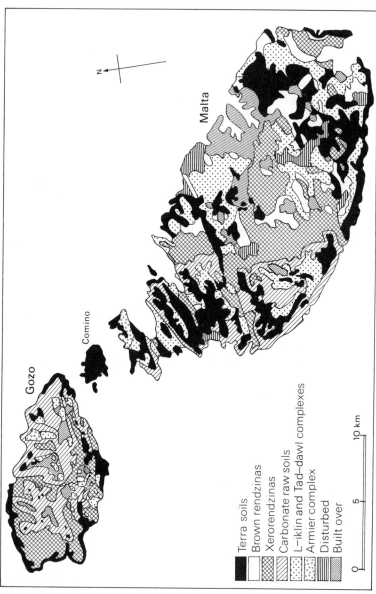

Fig. 79.—*Soils of the Maltese Islands.* (Based upon map in *Malta—Background for Development*, H. Bowen-Jones, J. C. Dewdney and W. B. Fisher.)

Gozo

Comino

Malta

Terra soils
Brown rendzinas
Xerorendzinas
Carbonate raw soils
L–iklin and Tad–dawl complexes
Armier complex
Disturbed
Built over

0   5   10 km

regions on a flat terrain, they are potentially much more fertile and may be heavily cultivated.

A Brief Summary of Maltese History

The earliest evidence of Maltese settlement is of Neolithic farmers in the third millennium B.C. Such evidence includes the so-called "cart-ruts" and some Megalithic rock tombs. An absence of abandoned pre-historic settlement sites may be accounted for by the suggestion that existing settlements have indeed been under continuous occupation since Phoenician and Roman times, but certainly there is little written and even less archaeological evidence of their occupations. From the ninth to the eleventh centuries, Malta appears to have languished under Islamic rule, and its return to Christendom was as a result of Norman assumption of control in Italy, Sicily, Malta and Gozo at the end of the eleventh century. Briefly passing through phases of Swabian and French control, Malta was passed to Aragon in 1283. Between this date and 1523, Malta remained a stagnant feudal backwater, and even came under threat from the Ottoman Turks during their expansion phase following the Conquest of Constantinople in 1453. However, in 1523 the Emperor Charles V of Austria, who was also Charles I of the Spains and their Aragonese Mediterranean Empire, bestowed the islands of Malta and Gozo upon the Hospitallers of the Order of St. John of Jerusalem, and this Order was to control the destiny of the islands down to 1798.

A commission appointed by the Order to report upon conditions in their new domain in 1524 described Malta as a "mere rock", inadequately covered with soil and with its main products being honey, cotton and cumin which the 12,000 inhabitants bartered for corn, water and timber which were said to be in perpetual scarcity. They also reported that the island was frequently subjected to the raids of corsairs. Gozo, with a population of 5,000, they described as more fertile. There were no ports whatsoever on Gozo, and the fine natural harbours of Valletta and elsewhere on Malta were neglected.

Over the period of the next 275 years the Knights lavished the wealth and care at their disposal upon fortifying castles, building churches and townhouses, and ensuring the well-being of the wealthy members of their order and the indigenous urban-dwelling nobility. Towards the rural population at large the Order maintained an "absent-minded paternalism", using wealth acquired elsewhere within the Order to keep the populace happy by providing them with cheap imported grain and other commodities, purchased in bulk and supplied to them at controlled prices. This served, along with the maintenance of domestic peace and political stability, to create for the Maltese a cocoon in which they were untouched by the changes fermenting in Europe at large. The urban patricians of Valletta adopted an Italianised outlook, culture and tongue in keeping with that of their patrons, the Knights of St. John. Meanwhile the rural population clung tenaciously to their subsistence agricultural way of life, and retained intact their ancient Semitic Maltese language. The same dichotomy persisted throughout the era of British rule which succeeded that of the Knights, and indeed it has been responsible for some of the present islanders' resistance to change and resentment and distrust of authority. The Maltese have been accustomed to living under paternalistic patronage, but the rural Maltese have grown to fear and distrust their urban patrons, and to cling to the security of their customary ways whenever possible.

Under British rule, Malta was declared a free port, and the islands became an entrepôt dependent upon the British presence. Under these circumstances agriculture and industry alike were neglected, and the adverse trade balance was made up from "services" for which Britain paid by providing secure government and economic subsidies to the islands, in much the same way as the Knights of St. John had done before them. In 1964 Malta acquired its independence within the British Commonwealth, but she has continued to depend upon heavy economic subsidies from overseas.

The Character of Maltese Agriculture

Agriculture provides employment for some 10 per cent of the total population of the islands, and occupies nearly two-thirds of the land area. Agricultural products also constitute a surprising 20 per cent of the country's exports. Although agriculture ranks so highly in the economy, this is some measure of the problems confronting the economy rather than an indication of prosperity. Some characteristics of Maltese agriculture may be summarised as follows.

First, the holdings are in general small and highly fragmented, and a high proportion of the land is rented. Entire holdings, which in more than half the cases are rather smaller than 9 tmien in extent (9 tmien = about 1 ha, *see* below), are frequently broken up into numerous tiny plots scattered over considerable areas—and rented from nearly as many landowners. It is, however, an ambition of many such small farmers to acquire at least part of their land for themselves.

Secondly, poverty is rife, because of small cash income, high land rents, and inefficient and uneconomic methods of farming. The problem of poverty is aggravated by a reluctance to invest the limited capital that may be available in holdings which are not owned and for which there is little security of tenure, particularly when faced additionally by insecure product markets. Because of a sense of foreboding farmers feel constrained to keep comparatively large sums of money in hand in expectation of the bad years to come, despite underlying ambitions to own land or build up livestock.

Thirdly, primitive customary techniques of cultivation are adhered to, and because of the small fields those farmers who have invested in mechanical aids have found that small rotivators serve their purpose better than large tractors, particularly since they require a smaller capital outlay than do the large machines.

Fourthly, illiteracy is widespread, particularly amongst the older farmers, who are consequently fearful and distrustful of government and "officialdom" generally. In the past they have all too easily fallen prey to the exploitation of the *pitkal* or middleman who handles the marketing of their produce. Consequently they cannot readily be persuaded to abandon old and "proved" methods in favour of modern techniques advocated by outside "experts", whom they regard either with distrust or as "very interesting" but not relevant to the Maltese situation.

Finally, the young people of the farming community are encouraged to obtain a good education, and many subsequently seek secure and remunerative employment away from the land, leaving an ageing rural farming population.

To help towards a fuller appreciation of the subsequent farm studies the following table of Maltese land measurements with their metric equivalents is included.

$$1 \text{ kejla} = 0.0018 \text{ ha}$$
$$10 \text{ kejliet} = 1 \text{ siegh} = 0.018 \text{ ha}$$
$$6 \text{ sieghan} = 1 \text{ tomna} = 0.108 \text{ ha}$$
$$9.3 \text{ tmien} = 1.0 \text{ ha}$$

A large-scale Gozitan livestock farmer specialising in the production of goat milk

Pawlu farms in the Gharb and San Lawrenz Districts of north-west Gozo, a region based upon Globigerina Limestone (*see* Fig. 80). He has built up his holding of 42.5 tmien (4.6 ha) by renting additional land, whenever this has become available, from a variety of landowners. The land, acquired through his own family and that of his bride, and through his own efforts since 1955, is rented from the Church authorities, the civil government, and other farmers and landowners over a very large area. In addition he has sub-let, sold, abandoned, declined an option on or exchanged a further 20 tmien, 3 sieghan and 8 kejliet (2.2 ha) on grounds of distance or inaccessibility. This still leaves him with land situated in twenty-six different locations, although he has managed to acquire 24 tmien (2.6 ha) in contiguous plots which makes the grazing of his goats so much easier. Pawlu actually owns the 4 sieghan or 0.67 of a tomna on which his farmhouse, cattle byre, goat shed,

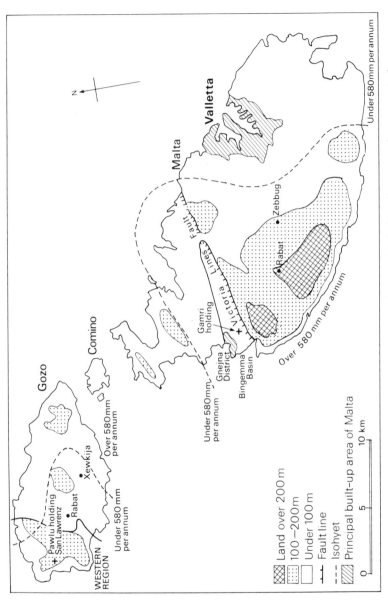

FIG. 80.—*The location of Maltese holdings referred to in the text, with further information relevant to agriculture.*

store rooms and a few small enclosures are situated. His land is utilised as follows.

	Tmien	Sieghan	Kejliet
House and adjacent roadway	0	2	0
Vines	0	3	0
Prickly pear	0	3	0
Potatoes	2	0	0
Vegetables, e.g. tomatoes and melons reduced from 6 tmien and 4 sieghan in 1956	0	4	0
Fodder crops, e.g. sulla, wheat, barley, mixed wheat and barley, and beans increased from 8 tmien 2 sieghan and 3 kejliet in 1956	14	2	3
Rough grazing and shooting	24	0	0
Unused	0	1	0
TOTAL HOLDING	42	3	3

Two crop rotations are employed:

(a) 1: wheat (after manuring and hoeing), 2: sulla, 3: wheat, and 4: fallow;

(b) 1: wheat, 2: sulla and beans, 3: potatoes, tomatoes and melons (after tractor ploughing), and 4: fallow.

NOTE: Sulla is a perennial legume, a nitrogenous plant of great value as a source of protein for livestock.

It has been Pawlu's policy to reduce vegetable production to what is required only for his family, on the grounds that there is no guaranteed market price for such produce, and to expand his production of otherwise costly fodder crops in order to sustain a larger herd of livestock, since the government pays him a guaranteed price for cow and goat milk, through its dairy at Xewkija. Here the milk is pasteurised before being marketed at cost, an indirect subsidy to the producer. Of the vegetables produced, most go to the family, a few are sold, while some vegetables, beans and prickly pear are used to feed the cattle and goats.

Livestock kept by Pawlu comprises 46 goats (including 8 kids) which produce the milk constituting the main source of income, 2 cows, 1 sheep kept for the production of "rikotta cheese" by Pawlu's wife, 100 head of poultry for egg and table-bird production, and a horse which is kept for draught purposes.

The production costs of Pawlu's holding include extra sulla, wheat and barley to supplement his own production, and in addition he makes purchases of supplementary cattle-cake. Pawlu owns no specialised farm equipment, but he is a member and organiser of a tractor co-operative through which the machines can be hired at well below the "going" commercial rate. Otherwise he depends upon his own horse and cart. He employs no outside labour, and himself works from the small hours of the morning until dusk, with only the help of a son in his middle teens, and casual help during school holidays from two student daughters and an older daughter who has a post as a teacher. Hired labour, according to Pawlu, is "out of the question", because charges made by such workers are so high that they often earn more than the farmer who employs them, who must bear in addition all the risks of production. Pawlu's wife also used to help on the holding, but now, apart from house and family, she only manages the care of lambs and chickens, egg collection, and the making of the cheese.

Compared to many other Maltese farmers, Pawlu is very progressive both in his attitude to his children—an older son and daughter have already acquired a good education with parental support and been lost to the land in consequence—and as regards his efforts to maximise the farm profitability by land consolidation, specialised commercial production, and support given to the tractor co-operative. On the other hand, like many Maltese farmers, Pawlu's marginal profitability might even be accounted a deficit if his labour were counted as a production cost, and his concentration upon milk production makes him wholly dependent upon existing government policy, particularly since he has diverted production from vegetables to fodder crops and pasture, rendering him particularly vulnerable to any change which might deprive him of this remunerative market.

An arable farmer in Malta with a diversified production

Gamri's entire holding, although fragmented into seventeen separate segments, is at least concentrated within an area of 2.6 km² within the Bingemma Basin and the Gnejna District of western Malta (*see* Fig. 80). The soils are mainly of Terra Rossa and Carbonate Raw varieties. The entire holding consists of 63 tmien (6.8 ha) rented by the farmer, including 53 tmien (5.7 ha) operated in partnership with his brother, and 10 tmien (just over 1 ha) managed by himself alone.

Of the total holding, 41 tmien (4.4 ha) are rented from private landlords, 9.6 tmien (just over 1 ha) from the Church, 6 tmien (0.65 ha) from the government, and 3.2 tmien (0.34 ha) from Gamri's father. There is one large plot of 21.5 tmien (2.3 ha) and another sixteen plots ranging from 1 to 8 tmien in extent, and the land is utilised as follows:

	Tmien	(ha)
Fruit (apples, citrus, etc.) and vines	21	(2.27)
Cereals and fodders, including barley, wheat, sulla, beans and greenstuffs	18	(1.94)
Land fallow, unproductive or sub-let	13	(1.40)
Potatoes	6	(0.65)
Other vegetables	5	(0.54)
TOTAL HOLDING	63	(6.80)

Rotation includes the following variants.

On good land:
  (a) 1. potatoes *or* winter fallow, followed by a summer crop such
         as potatoes *or* melons;
      2. potatoes *or* tomatoes;
      3. sulla;
      4. wheat.

On poor land:
  (b) 1. wheat *or* barley;
      2. sulla or vetches.

On the best irrigated land:
  (c) 1. winter-sown potatoes, followed by autumn-sown potatoes;
or (d) 1. winter-sown potatoes;
      2. tomatoes *or* melons;
or (e) 1. land irrigated for potatoes *or* tomatoes *or* melons in one year;
      2. cultivated "dry" in the subsequent year for cereals, beans and
         onions.

Apples are the most important "cash-crop", and are grown in a
sheltered valley along with a variety of other fruit such as lemons,
oranges, mulberries, figs, cherries and plums. There is a problem of
boring insects which infest the trunks of apple trees, and the brothers
waste much time each autumn gouging the larvae out with penknives
and pieces of wire, because they are dubious about the efficacy of pesti-
cides. This is just one example of the reluctance of such farmers to in-
novate for fear of disaster. Fruit contributes 72.6 per cent of the total
joint income of the brothers.

Grapes are regarded as especially profitable, in particular wine grapes
because these require less attention to quality. The vineyards are
worked over three times with hoes in late winter and early spring, and
about 25 kg of ammonium sulphate per tomna (233 kg per ha) is worked
into the soil at the second cultivation. The grapes produced are worth
another 12 per cent of the joint income.

Wheat and barley are sown using seed from a previous crop. The
land is worked over twice in the previous autumn with a wooden

plough. The seed and fertiliser are applied together by broadcasting at the rate of 25–50 kg per tomna, or 233–465 kg per ha, and ploughed in immediately afterwards. Sulla seed is also derived from the previous crop. Although sulla cultivation involves a considerable outlay in labour and land, it also contributes substantially to the fodder requirements of the holding, along with the barley, wheat and beans. Furthermore, surplus sulla sold to other farmers contributes an additional 6 per cent to the total joint income.

Seed potatoes are an imported British variety, e.g. Arran Banner, and this crop takes its place in several rotations. Even in a poor year potatoes contribute a further 2.6 per cent to joint income. Either melons or tomatoes are included in the main rotation, as well as that of the best irrigated land. Both crops require a lot of ploughing, but melons contribute 2.3 per cent and tomatoes 4.3 per cent to the total income of the holding.

Although the farm income, as demonstrated above, is derived wholly from arable production, Gamri and his brother have been building up a herd of pigs which they hope will enable them to achieve a greater stability of income than that derived solely from sale of crops. Accordingly there are at present 5 sows and their litters, about 40 pigs in all. In addition there is one mule kept for ploughing and carting around the holding, 2 goats and 2 sheep which supply most of the families' milk requirements, and some 24 head of poultry kept for their meat and eggs. A number of rabbits are also reared for the table, and additional meat is obtained by shooting as well as fishing.

It is most revealing to consider the contribution of each crop to farm income, in relation to the inputs of land and labour it requires. It is clear that fruit and vine cultivation, while they absorb 34 per cent of the land resource and 39 per cent of labour, also provide 85 per cent of gross income, and are therefore quite the most economic activity carried on. Fodder crops occupy 29 per cent of the farm area and demand 24 per cent of the labour available, yet their cash contribution is confined to the 6 per cent derived from sales of sulla. It must, however, be borne in mind that these crops sustain the mule which provides the farm with draught, provide food for both families, and feed pigs which are being reared as an investment for the future. Finally, potatoes and vegetables occupy 18 per cent of the land area and utilise 21 per cent of labour; yet, in addition to providing part of the two families' staple diet, they yield 9 per cent of the joint farm income.

More than 15 per cent of the total man-hours devoted to the holding contributes no tangible income, yet a large proportion of these 1,046 man-hours may best be seen as providing a substitute for capital input, which is negligible. Land input too is negligible, since, of the total land area, 20 per cent at any given time is non-productive, leaving only an effective productive area of 50 tmien (5.4 ha).

It could be argued, employing the criteria of the commercial farmer, that a holding of 5.6 ha which, having absorbed 6,816 man-hours of labour, can only yield the equivalent of £468 joint income, is wholly uneconomic. This, however, is totally to disregard the tendency of Maltese farming people, although motivated towards a cash profit, still to adhere to an older subsistence way of life. The lavish expenditure in man-hours can therefore be justified in much the way expressed by the authors of *The Kibbutz Experience* at the end of the Kibbutz Kfar Blum study in Chapter XIV, on p. 243. The input of about 110 man-hours per tomna (982 per ha) is then seen as providing sustenance for Gamri, his wife and eight children, his brother and his wife, their three sons and a grown-up nephew, a total of five adults and eleven children supported with a minimum of capital or land. If this is assessed in money terms an entirely different interpretation of the holding's viability is evident.

Chapter XII

Intensive Rotational Mixed Farming in Western Europe

Introduction: Evolution of the System

THE following factors operative in Western Europe have contributed to the evolution of the system of intensive rotational mixed farming.

Intensity of cultivation

Industrialisation, which began in this region, has created large urban populations wholly dependent upon agriculture for essential foodstuffs and raw materials, while at the same time urban expansion has reduced the area of agricultural land available. To meet the growing demands made upon a diminishing area of productive land, increased applications of fertiliser have been employed to make heavier cropping possible, and mechanical aids have been employed to make up for the manpower lost to industry. Intensive use of crop rotations, fertilisers and mechanisation have all contributed to some of the largest crop yields per area of land ever achieved.

Development of sophisticated crop rotations

Chapter VII discussed the evolution of crop rotation. Primitive rotations provided early farmers with a means of sustaining land fertility when shortage of cultivable area obliged them to crop their holdings continuously. At first a rest year or fallow had to be included in such rotations, but the introduction of improved fertilisers and new crops by which rotations could be diversified has so diminished the threat of soil exhaustion that continuous cropping without fallow, yet without detriment to soil fertility, is now feasible (*see* p. 213).

Mixed farming

Cultivation of a variety of crops and/or the keeping of several varieties of livestock on a single holding have, like rotational practices, had a long history in this region. In the Middle Ages the Saxons in northern Germany, the Jutes in the Jutland Peninsula and the Anglo-Saxons in England developed diverse arable and livestock economies to minimise the risks of starvation that could be brought about by crop failure in the unpredictable cyclone-dominated maritime climate, since food shortages could not be alleviated by trade because of the inadequate communications in that period. The "mixed" economy favoured self-sufficiency, because whatever the summer conditions some crop or animal might be able to prosper, hence the Anglo-Saxon maxim, "Down horn—up corn", for when cattle suffered from arid pasture, a good wheat harvest was assured, and vice versa.

Later the emergence of commerce and industry encouraged Western European farmers to switch from subsistence to commercial production. The new dependence of farmers on their industrial markets rendered them for the first time susceptible to those same fluctuations in the level of economic activity—"booms and slumps"—to which industry so often is subject. Diversification of production has, however, placed the European mixed farmer in a good position, because of the greater flexibility of output he can employ in order to offset changes in market demand.

To illustrate some regional diversities that occur within the system, examples are cited from the Netherlands, western Norway and the United Kingdom.

A General Description of Netherlands Agriculture

Much of the Netherlands is made up of drift and recent deposits, many of which have been laid down by the sea, by rivers—particularly the various distributaries of the Rhine—and by the Great Continental Ice Sheet. A notable exception to this is the chalk formation which underlies Limburg in the extreme south of the country, much of which is covered by a mantle of loess, a dust of aeolian (wind-borne) origins redeposited by the Quaternary ice sheet. The main soils of the Netherlands and their agricultural uses are illustrated in Fig. 81.

The climate of the Netherlands is homogeneous, since the country is small and flat and everywhere subject both to the mildening influences of the sea and to the unpredictable conditions which characterise cyclonic dominance. The rainfall is well distributed, displaying two maxima, one in July–August, characterised by rain of convectional origin, and one in September–November, characterised by cyclonic rain. Temperature and rainfall characteristics are summarised in the following statistics.

January mean temperature 1.9°C
July mean temperature 16.8°C
Annual temperature range: 14.9°C
Rainfall:
    spring quarter (March–May): 140 mm
    summer quarter (June–August): 216 mm
    autumn quarter (September–November): 219 mm
    winter quarter (December–February): 180 mm
    Total annual rainfall: 755 mm

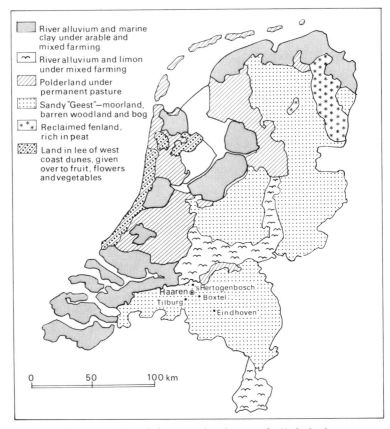

River alluvium and marine
clay under arable and
mixed farming

River alluvium and limon
under mixed farming

Polderland under
permanent pasture

Sandy "Geest"—moorland,
barren woodland and bog

Reclaimed fenland,
rich in peat

Land in lee of west
coast dunes, given
over to fruit, flowers
and vegetables

'sHertogenbosch
Haaren
Boxtel
Tilburg
Eindhoven

0   50   100 km

FIG. 81.—*Soils and their agricultural uses in the Netherlands.*

Agriculture in the Netherlands has benefited greatly from two significant past developments: first, the land reclamation process which has added some 40 per cent to the country's total land area since medieval times, mainly around the IJsselmeer; and secondly, a large expansion of urban industrial markets for agricultural products, as a result of which agriculture in the Netherlands has experienced growth in market demand as well as productive area. These factors have contributed to a vast growth in the scale of agricultural production during the current century.

The total land area of the Netherlands is 41,160 km^2, and the population stood at 13.5 million in 1974. The government conducts regular and exhaustive censuses covering almost every aspect of agriculture through its Central Bureau of Statistics, and the data for a given year are published in full by the Department of Agriculture in June of the following year. The annual census in May is primarily concerned with land areas, crops and livestock statistics. There is also a quinquennial census concerned with details of the agricultural labour force, land-holdings and ownership, the maturity of orchards, and the state and nature of mechanisation.

In 1974 the total cultivated area was 2,091,585 ha, and there were 166,197 agricultural holdings. This would represent an average agricultural holding of about 12.5 ha. The total number of holdings includes,

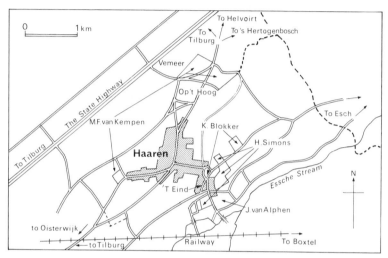

FIG. 82.—*Location of farms in the Haaren district of North Brabant.* M. F. Van Kempen and K. Blokker are discussed in Chapter XIII.

of course, the numerous, tiny horticultural holdings which characterise the country (*see* Chapter XIII). Land-use statistics indicate that about 56 per cent of the total productive agricultural area is in permanent grass, 38.5 per cent in arable crops, 5 per cent in market-garden produce, and 0.5 per cent in bulbs and flowers. Agricultural products make up some 25 per cent of the country's exports.

To illustrate intensive rotational mixed farming in the Netherlands, which will subsequently be compared with that of western Norway and the United Kingdom, the following farm-studies are taken from around the North Brabant town of Haaren, in the southern and Catholic sector of the country (*see* Figs. 81 and 82). In this region most of the land is about 5 m above sea-level, and the soil is fine and sandy.

Farm-studies around Haaren, North Brabant

The Vemeer Farm at Hoge Raam, north of Haaren

This holding, containing 40 ha of land, is made up of what was originally two separate farms. The farm is situated in flat, open country which is regarded as an important conservation area. The home section, and some other parts of the farm, have a loamy soil with a surface layer of about 1 m of humus-rich black soil. Several other non-contiguous plots have 1 m of sandy surface soil. The drainage of these soils is good. By amalgamating the smaller fields whenever possible, better drainage and land-use has been achieved. The joining of the two farms has made possible more effective use of machinery and better operation generally. This farm is not wholly typical of the region, since most farmers keep fewer cows, and pigs are a more important part of their farm economy.

Livestock on the Vemeer Farm includes 170 milk cows and a bull, of the Maas-Rhine and IJssel breed. To provide sufficient fodder on so small a holding, very intensive cultivation must be practised, and rotational grass, maize, beet and peas are grown. The calves and old cows are marketed for beef.

In spring, ploughing is followed by a liberal application of artificial manure, and the sowing of grass, maize and beet. In autumn the crops are harvested and used to make silage. The 26 ha of rotational grass yields 7,000 kg of silage per annum; 2 ha of beet ("Corona" or "Majoral") yield 10,000 kg; finally, the 12 ha of maize give a yield of about 7,000 kg per annum. All these crops go to provide fodder for the cattle.

There are two farmhouses on the property, plus a modern dairy and several other barns. The two original farm-owners work together in partnership on the holding. One full-time tractor driver and one part-time farmworker are employed, and occasionally students are employed during the peak summer season.

The tractors, one old, one quite new, are kept exclusively for work on the holding, since some 2,000 hours of tractor work are required on the holding each year. The cattle go to market by tractor and trailer, and the tractors also provide for internal freight movement and for field operations.

The sole source of income on this farm is cattle.

The Op 't Hoog Farm, Helvoirtseweg, north of Haaren

The Helvoirt Road, in the north of the town of Haaren, branches off to 's Hertogenbosch and Tilburg, and the Op 't Hoog Farm is situated nearby. This farm is composed of 11 ha of owned land and 9 ha of rented land, making a total of 20 ha. This is primarily a cattle and pig holding on which most of the land is in grass, although about 5 ha are usually devoted to maize.

There is a breeding unit for about 90 sows and 3 boars which measures 60×9 m. In 1974 a new fattening pen 47×14 m was constructed with a capacity for 500 fat-pigs. In 1976 a new cattle house was constructed to accommodate 60 milking cows and measuring 38×16 m.

The sandy soil of the holding has 50 mm of black humus topsoil which enhances its fertility. Twenty years ago the holding was rather smaller, and a larger proportion was given over to crops such as oats, rye, barley, potatoes and beet, which left pasturage for only 7–8 cows compared with the 60 milk cows and some 500 fat-pigs which the holding can support today.

Although privately owned and operated, the farm buys 500,000 kg of fodder per annum through co-operative wholesalers, and sells its fat-pigs through a co-operative abattoir and its milk through a co-operative dairy. The cows are milked in a modern, fully automated herring-bone dairy of 12×7 m.

When harvested the grass and maize are both ensilaged by a man who undertakes this work under contract on behalf of all the local farmers. In winter the silage can be extracted for the livestock using a silage-cutter.

All regular farm labour is furnished by father and son. They are able to handle the operation themselves by employing machinery wherever possible, such as tractor, plough, cultivator, chain harrow, hay trailer, slurry tanker, silage-cutter, hay maker, cyclo-mower, a spreader for artificial fertilisers, and a number of trailers for internal transport or for cattle despatch.

The farm income is derived 50 per cent from cattle and milk, and 50 per cent from the pigs.

H. Simons's farm, Eind, south-east of Haaren

This farm (*see* Fig. 83) consists of 16 ha of land, including 13 ha of owned land and another 3 ha of rented land. Of this, 9 ha are in a single open field which makes it suitable for a wide range of uses. There are a further 7 ha inside the Nemelaer Nature Reserve, upon which certain land-use restrictions are imposed.

The soil is inclined towards dryness, being well drained and sandy. The natural vegetation, as found still within the Nature Reserve, includes oak, alder and poplar trees, rushes and grass, which contrasts with the evident improvements in the arable lands achieved over the past hundred years by means of land reclamation, improved drainage, use of fertilisers and reformed land-holdings.

De Heer Simons has achieved economies of operation by membership of several agricultural co-operatives, one of which provides him with fodder and fertilisers, while he also sells his milk to a co-operative dairy. In all other respects he is an independent proprietor, and typical

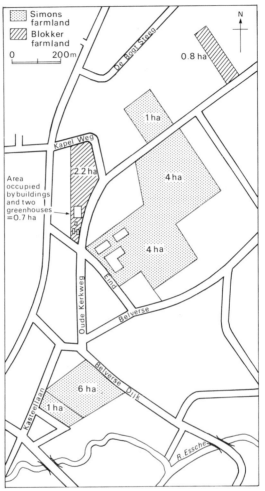

FIG. 83.—*H. Simons's and K. Blokker's farms.* For K. Blokker, *see* Chapter XIII.

of the region. Livestock consists primarily of a herd of 80 cows of the Maas-Rhine and IJssel breed (a dairy breed), and horses, dogs, ducks and rabbits which are kept for pleasure. To get the cows in milk they are impregnated by artificial insemination, and this ensures a higher quality in the "followers" produced for beef than would be achieved by keeping a bull. As far as possible the cattle are fed on home-produced silage and maize.

Seasonal work is minimal; 10 ha of sown grass (a mixture which includes English and Italian varieties, field-grass, timothy and clover) and some 6 ha of the Nemelaer, which are given over to maize, are planted in the spring. Silage-making is begun in May and the maize is harvested in late September. The pastures yield 1,500,000 kg, or 500,000 kg three times yearly, for silage, while the maize yields some 50 tonnes of fodder.

The farmhouse covers an area of 15 × 10 m (*see* Fig. 84). The dairy is equipped with modern electric milking machines and a cooler with

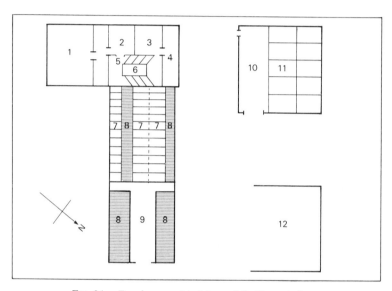

FIG. 84.—*Farmhouse and buildings of H. Simons's farm.*

Key

1. Dwelling house—private	7. Stalls for cattle
2. Shed	8. Gratings for dung storage
3. Milk cooling tank	9. Storage for fodder
4. Roofed area	10. Fodder stores for calves
5. Cattle waiting area	11. Calves' pen
6. Milk well	12. Machine store

capacity for 3,000 litres of milk. The whole operation can be managed solely by the farmer with only some spare-time help from a son.

Mechanical aids include a tractor, a two-furrow plough, a hay-tedder for fluffing up the hay in the field, a mower, a spreader for synthetic fertilisers, a silage-cutter and a number of trailers for transport within the farm or the marketing of livestock. None of these pieces of equipment are put out to hire.

The co-operative dairy collects the milk from the farm daily in a mammoth tanker with a capacity of some 20,000 litres.

Farm income is derived 70 per cent from milk and 30 per cent from meat.

J. van Alphen's farm at Driehoekweg, south-east of Haaren

This is a 12 ha holding, which includes arable and pasture land, as well as a large number of specialised buildings in addition to the 29 × 14 m farmhouse, and a shed of 15 m × 6.5 m. These buildings include seven pig-pens ranging in size from two with dimensions of 34.35 × 8.55 m down to one of 15 × 4.5 m.

The land is flat with a sandy loam soil. The holding is an entirely independent operation although a slurry tank is maintained and operated in conjunction with two other farmers.

This farm is typical of its region in most respects. There are 1,000 pigs, roughly equal in the numbers of male and female, which have been purchased from contributory stockbreeders at a pre-agreed number and price. The young pigs arrive weighing about 22 kg, and the males, which are not required for breeding, are promptly castrated in order to facilitate fattening. Four months later they have reached a slaughter-weight of about 100 kg.

Fodder is purchased through the Agricultural Wholesale Co-operative Society. The Co-operative Abattoir at Boxtel handles about 21,000 pigs per week. After the pigs have been sold the pig-pens are fumigated in readiness for a new batch.

In spring the arable area is ploughed and sown with maize for sale to cattle farmers in need of fodder for the subsequent winter. In winter a dressing of slurry helps maintain soil fertility.

The pig-pens all follow a standard pattern, with an insulated roof and an interior divided into stalls each of which accommodate some 8–10 pigs. Three pigs will eat at a trough. The individual stalls have slatted floors: 2.2 m covered with bedding, and 1.15 m of grating through which dung falls to be accumulated in a dung-cellar below, following the Danish model.

Labour on the farm is provided by the farmer unaided, and he uses a tractor, two ploughs and a cultivator, all solely for his own work. The slurry tank, which has a capacity of 4,000 litres, is shared with two other farmers.

The farm does not require any transport since the pigs are collected from the farm by the abattoir, and purchasers of maize also collect it direct from the farm.

Farm income is derived 80 per cent from the fat-pigs and 20 per cent from maize.

Rosendal in Kvinnherad Kommune, Western Norway

Rosendal is a small town with a population of about 1,000, situated almost at the intersection of long. 6° E. and lat. 60° N. on the southerly

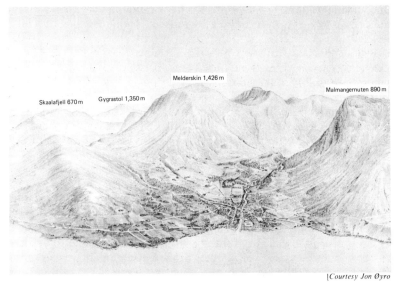

FIG. 85.—*Bird's-eye view of Rosendal.*

[*Courtesy Jon Øyro*

shore of Hardangerfjord. It is the principal township of the Kvinnherad Kommune or Rural District, a region of about 70 km north to south by 30 km east to west. The attractive town is fringed on the west by Hardangerfjord, with a frame of lofty mountains on the other three sides (*see* Fig. 85). From glacial tarns high in these mountains torrents descend to form the Mel and Hatteberg Rivers, which in turn merge some 500 m from the shore. According to some authorities it is from the twinning of these rivers that Kvinnherad derives its name.

Factors which determined the agricultural character of the region

Bedrock consists of pre-Cambrian igneous rocks, overlain by Cambro-Silurian rocks of various kinds, notably highly metamorphosed crystalline rocks such as greywacke, limestone and schists, but displaying a number of igneous rocks, intrusive and extrusive. The metamorphism was a consequence of pressures imposed during the Caledonian Orogeny (mountain-building movement) some 400 million years ago. It was then that the Scandinavian Mountains acquired their characteristic north-north-east, south-south-west fold-trend, similar to that displayed in the north-west Highlands of Scotland. In the Rosendal district this alignment is reflected in the shape of Hardangerfjord and those of the off-shore islands of Snilstveitoy and Scorpo.

Another determinant of relief and drainage patterns has been the continental ice sheet, which subjected western Norway to severe erosion in course of its passage westwards from the mountain crests to the sea. Successive recessions are marked by terminal moraines, one of which lies across Hardangerfjord to the south-west of Rosendal, others to the east and north-east of the town. The grandeur of this breathtaking glacial scenery has greatly enhanced the tourist potential of the region, while the same glacial erosion has made possible a rich resource of hydroelectric power. Over the years mountain torrents cascading down the steep mountain flanks have borne glacial debris towards the lowlands below, and rivers like the Mel and Hatteberg have deposited this debris on the floodplain, and carried it out into the less-turbulent fjord waters to form deltas. One such delta is that situated at Rosendal, deposited by the common waters of the Mel and Hatteberg, forming a flat area something like 1 km² in extent, which once was a major agricultural resource. The delta is composed of fluvio-glacial deposits graded seawards from coarse round pebbles down to a fine sand, but overlain by a mantle of fertile fluvio-glacial alluvium about 110–380 mm thick. Here, because the soil is naturally porous, fields are liable to dry out in hot summer weather. Today much of the delta land is no longer farmland; instead a secondary school, a hotel and several houses have been established near the shore. Further inland, where the farm-studies subsequently described were carried out, the soil is composed of a heavy, water-retentive, morainic boulder clay, covered by the debris of successive landslides from adjacent mountain slopes.

The land rises slowly inland from the shoreline, but from 100 m above sea-level the mountain slopes ascend steeply, clothed in mixed woodland with deciduous trees such as birch, alder, ash, European cherry, rowan, hazel, aspen, willow and elm predominant. Only further north are conifers predominant, although where natural woodland has been cleared they have been replaced by stands of commercial spruce. It is indicative of the comparatively mild climate in this region that, despite the high latitude, the tree zone extends to 790 m. Above this the upland fjelds are clothed in grasses, lichens, mosses and several varieties of ling and heather. Yet higher, at some 1,350 m the permanent snow-line is reached, about 10 km east of Rosendal, above which Norway's third largest glacier, the Folgefonn, broods upon the mountain heights.

The mild, wet climate of western Norway, milder than might be expected in these high latitudes, has been a major determinant of the natural vegetation. The equable temperatures may be attributed to such factors as:

1. the different cooling rates of land and sea which render maritime land areas slow to cool in winter, but also slow to warm in summer;
2. the winter warmth brought along this coast by the North Atlantic Drift, the northern extension of the Gulf Stream;

3. the comparatively sheltered situation of Rosendal which shields it from the cold easterly winds which prevail in winter.

The heavy rainfall is attributable both to its position athwart the most frequented winter cyclonic track, and to the nearby mountains which induce heavy orographic rain in autumn and early winter. Temperature and rainfall statistics are as follows.

Jan. mean temperature	July mean temperature	Annual temperature range
1°C	15°C	14°C
Total annual rainfall	Oct. maximum rainfall	May, minimum rainfall
1830 mm	205 mm	68 mm

Prevailing winds are from the south-west and south-east over much of the year. It is, however, the south-east winds which, when they occur in winter, blowing out from the continental high-pressure interior, and accentuated by the natural valleys which occur to the east of Rosendal, may cause severe damage to property and large trees. This, however, is unusual, and for the most part the climate is mild for reasons already described. This has favoured the agricultural development of the region.

Agriculture in the Rosendal district

Years ago agriculture was the main source of income in this district, with subsidiary fishing and craft industries. While it is still important, several new occupations have emerged, and some older ones have acquired a new importance. New occupations include aluminium-smelting, a dairy, several hydroelectric plants, and new service industries related to transport and tourism. Whereas fishing and furniture-making have declined, boat and ship building have developed. Agriculture seemed at one time set in decline as new industries began to attract its labour force away, offering wages and conditions with which the small farms could not hope to compete. Under economic pressures, small farm units have indeed given way to larger units made up by amalgamation, better able, by greater economy of production, to withstand such pressures. In recent years the Norwegian Government has adopted policies designed to enable the agricultural sector to compete more effectively for available labour, and has promoted agricultural research towards more cost-effective farm production.

Because of the over-all limitations imposed by the rugged relief upon the total farm-land available, even such new consolidated holdings are rather small by some Western standards. This is reflected in the adoption of a small metric unit of land measurement, the dekar, which is only one-tenth the size of the widely accepted hectare.

By careful modernisation and rationalisation on the farms, milk output has been so increased that growing demand for fresh milk locally
and further afield can be met, and a remaining surplus used for the
production of cheese.

Cheese factory at Seimsfoss

The Kvinnherad Ysteri (cheese factory) at Seimsfoss, 3 km south of
Rosendal (*see* Fig. 86), receives the milk production of about half of
Kvinnherad, the area which includes Rosendal. At the time of the study
about 6,500,000 litres of milk a year were supplied by 112 producers,
not counting butter received from eight farmers in isolated districts.
The farmers received 1.84 kr per kg of milk, less 0.09 kr per kg deducted
by the Central Dairies in Bergen to cover administrative and transport
costs. The milk is collected from the farms in mammoth tankers, which
pump the milk from cooling tanks installed on all the contributory
farms. Milk production remains comparatively stable despite a slight
decline in the number of producers. There is a large demand for fresh
milk in Norway with per capita consumption standing at 220 litres per
annum. Consequently, only the balance of the milk collected at Seimsfoss remains available for the production of cheese. Out of a total production and sales income of 2.2 million kr. in 1977, 1.7 million kr. was
in milk. Most of the milk was despatched to Bergen.

[*Courtesy Gunnar Øyro*

Fig. 86.—*The Kvinnherad Ysteri* (*cheese factory*) *at Seimsfoss*. The factory receives milk
from the whole of Kvinnherad, makes cheese, packs milk for consumers and delivers
to shops in the area.

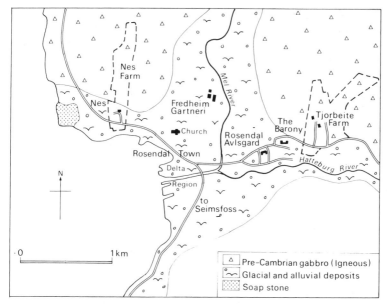

FIG. 87.—*Rosendal: location of features described in text*. (Fredheim Gartneri is described in Chapter XIII.)

The dairy produces a traditional Norwegian cheese known as *pultost*, which is made from sour and skimmed milk, and for which output in 1977 stood at about 65 tonnes, some of which was sold to local shops, while some was sold through the Bergen Section of the Norwegian Dairy Marketing Agency.

The dairy also received 2 tonnes of red and black currants and raspberries which were despatched to jam and juice factories in Bergen. As an extra service to its customers, the dairy also cooked 68 tonnes of potatoes for fodder in 1967.

The labour force consists of six full-time male workers and three part-time female workers.

Rosendal's farms

Most of Rosendal's present-day farms are named after and derived from the ancient medieval farms of Malmanger, Veng, Skala, Nes, Mel, Om, Seim and Hatteberg. Two farms subsequently described are Nes and Tjorbeite, the latter a part of Hatteberg (*see* Fig. 87).

The farms of Seim, Mel and Hatteberg were originally included in the estates of Baron Ludwig Rosenkrantz, who in the seventeenth century administered them from his handsome baroque mansion situated at the foot of Melderskin and still known today as "the Barony". Subsequently, when the estates were broken up, the mansion passed to Oslo

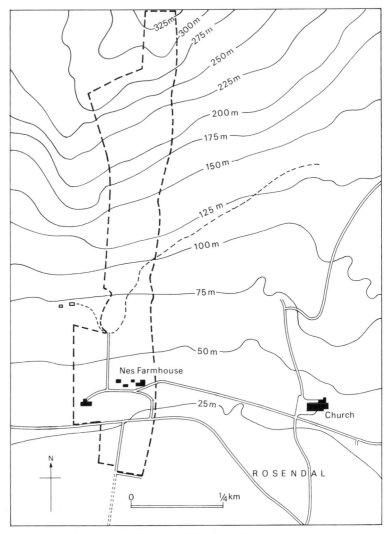

FIG. 88.—*Map of Nes Farm, Rosendal.*

University, and one farm was developed into an agricultural research establishment under the control of the Agricultural College of Norway (*see* below).

The agricultural research establishment in Rosendal

The Avlsgård Farm of Rosendal Barony has been the location of a series of experiments in sheep husbandry intended to provide new crossbreeds which will prove more productive and therefore more economic for Norwegian farmers. To promote stock improvement more than 1,000 animals from throughout the region of Kvinnherad are exhibited here each autumn.

One experiment which was undertaken was to ascertain the advantages, if any, for stock improvement of three-breed crossing of sheep over two-breed crossing or breeding pure stock. This experiment, completed in 1968, proved of little significance for the breeds involved, or for the traits considered. Another series of experiments conducted about that time were directed towards the development of ewes capable of multiple suckling in addition to multiple birth. Further experiments are proceeding at the present time.

A study of Nes farm

Nes is situated just north of Rosendal, near the "nose" or peninsula from which it derives its name. The farm has been made up from two smaller holdings out of the four parcels of land into which the medieval farm was at first partitioned. The present farm consists of 376 dekar (37.6 ha) (*see* Figs. 88 and 89).

At this point the base-rock is gabbro, the subsoil a heavy, impervious boulder clay. Consequently, although the surface soil is very fertile, the farm-lands have to be drained with ditches and pipes. The soil is rich in sulphur and potassium, with a pH value of about 6.0, slightly acid.

[*Courtesy Gunnar Øyro*

Fig. 89.—*Olav Nes's farm and land.*

Periodically the land is dressed with calcareous seá-sand, derived from shell fragments, to reduce acidity, or calcium-rich fertilisers are used. In spring, once the soil is sufficiently dry, usually in April or early May, it is ploughed and harrowed and its composition balanced by judicious use of manure and Fullgjødsel 19–5–9. The numerals denote the percentage content of nitrate, phosphorus and potassium. The fertiliser also contributes sulphur, chlorine and magnesium to the soil. Another fertiliser, free of chlorine, is used in the production of potatoes, tomatoes and strawberries.

The oldest records appertaining to Nes date from 1405, but several tumuli, or burial mounds, and various artefacts testify to Bronze or

Fig. 90.—*Olav Nes's farmhouse and adjacent buildings.*

even Stone Age settlements in the area. The present farmhouse probably dates from about 1860–70, a period when Norwegian land-holding reforms were taking place. An older pattern of nucleated farm villages or "tuns" comprising 8–10 families, with 50–100 buildings, and surrounded by their village lands, gave way at this time to one of isolated farmhouses on new consolidated holdings. The farmers of that day either moved their older wooden farmhouses piecemeal to the new site, or, as the Nes farmer did, built a new farmhouse there (*see* Fig. 90). The barn and cattle sheds, however, were built in 1970, are of modern design, and are situated barely 70 m from the road in order to facilitate milk collection by the dairy at Seimsfoss.

Even in the early twentieth century, farmers in western Norway were operating on a subsistence basis. They kept cows, sheep, a few pigs and one draught horse and grew crops of barley, oats, rye, wheat, potatoes and a wide variety of vegetables as well as grass, of which as much as possible was converted to hay for winter use. There was insufficient winter fodder for livestock unless the sheep, but not the cattle, were driven up to the saeters (summer pastures) for grazing each summer. Since the Second World War this traditional practice has been rendered unnecessary by improvement of fertilisers and cultivation techniques which ensure that there is sufficient fodder for all the livestock, despite an over-all increase in their numbers. In 1970 Herr Nes had 35 sheep, but discontinued keeping sheep in 1976. He also no longer keeps any pigs, and horses have not been kept on the farm for some years.

In August 1978 there were 21 cows, 1 steer (a number having already been marketed), 6 heifers and 12 cow calves for herd replacement. These numbers, compared with 11 cows and 15 "followers" in 1970, are a

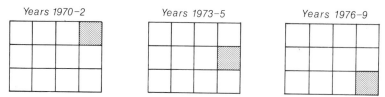

FIG. 91.—*Crop rotation employed on 60 dekar (6 ha) arable field on Nes Farm.* Shaded portion represents 5 dekar (0.5 ha) of which 2 dekar are devoted to potatoes and 3 to turnips for successive periods of three years, while the remaining 55 dekar are put into rotational grass.

measure of the increased size of herd made possible by improved fodder production. The cattle are Norwegian Red Cattle, a breed suitable for milk and beef production. Some calves are also reared for sale as breeding stock to other farmers.

In 1970, 50 dekar of arable land were utilised as follows: 36 of grass for silage, 7 of potatoes, 4 of barley, and 3 of root crops for fodder. Now, of 90 dekar of arable land, 30 are rented and are used entirely for grass. The remaining 60 dekar are farmed as a single large field employing a form of rotation, with 55 dekar of grass, 3 of potatoes and 2 of turnips. After three years in one location, potatoes and turnips are moved to a new location for the subsequent three (*see* Fig. 91). The grass is a mixture of timothy, clover, rye grass, blue grass and meadow fescue, and is used mainly for silage but some for hay. The potatoes (Pimpernel, Vestar and Kerr's Pink) yield 3,000 kg per dekar, and half goes for human consumption, the rest for fodder. The turnips (Swedish Yellow) yield 7,000 kg per dekar, and all but 200 kg are used for fodder. Sowing is in April and May, and pests and weeds are controlled by spraying by tractor in early summer.

TABLE 10

Report by the Norwegian Agricultural Economics Institute for
Olav L. Nes's farm, 1976

	Olav Nes's farm	All farms average	Technically advanced farms (average)
ARABLE PRODUCTION (DEKAR)			
Fully cultivated	116	135	139
Rented	30	26	31
Potatoes	3	2	2
Sown grass and permanent pasture	114	131	135
CROP YIELDS (KG PER DEKAR)			
Potatoes	1,500	2,403	2,408
Grass and pasture	443	404	450
PRODUCTION INCOME (KR.)			
Potatoes	1,860	870	950
Other crops	3,845	3,042	3,908
PRODUCTION COSTS (KR.)			
Seed (e.g. turnips)	1,175	559	575
Fertilisers	7,606	9,818	11,467
Conservation means	3,400	2,745	3,187
HOURS FARMWORK (PER ANNUM)	4,342	4,249	4,609
LIVESTOCK			
Cattle (average number)	20.4	17.7	19.1
Milk per cow (kg)	5,246	5,489	5,780
Price of milk per kg (kr.)	1.70	1.76	1.76
Income from milk (kr.)	170,037	161,852	184,354
Income from beef (kr.)	46,432	49,826	59,464
Sheep (number)	26	5	5
Price of mutton per kg (kr.)	20.81	20.77	21.75
Income from mutton (kr.)	16,735	2,512	2,814
Income from wool (kr.)	1,683	265	278
SUMMARY OF INCOME (KR.)			
Animal products	234,887	215,590	248,283
Arable products	5,705	3,912	4,857
Government subsidies	29,776	29,284	31,556
Other sources	1,983	2,337	3,071
Total from all sources	272,351	251,123	287,767

Besides the 90 dekar of arable land, there are 26 dekar of improved pasture, 160 dekar of unimproved pasture and 100 dekar of woodland. The woodland contains some birch fit only for firewood, and young spruce not yet mature enough for commercial exploitation.

Herr Nes owns a tractor, harrow, plough, hay-maker and grass-mower. Other machines, for example manure and fertiliser spreaders and land-rollers, he rents through a "machine ring" shared between him and several other farmers. These machines enable him to produce more foodstuffs at lower cost, with only the limited labour afforded by his own family, since outside labour is in short supply and costly. The entire farm is operated by Herr Nes and his wife, with spare-time help from a schoolboy son and 21-year-old son in full employment elsewhere, and some help from his able-bodied 83-year-old father and young daughter of 13.

Herr Nes is a member of the Bergen Central Dairy, the Norwegian Meat and Pork Centre (N.K.F.), the West Norwegian Farmers' Produce Marketing Association, and the Norwegian Red Cattle Association (N.R.F.). All these associations enable him and other small farmers

TABLE 11

Detailed breakdown of Olav L. Nes's income for 1976

Source	%
Beef	17.0
Milk	62.5
Mutton	6.1
Wool	0.6
Potatoes	0.7
Other crops	1.4
Government subsidies	11.0
Other sources	0.7
TOTAL	100.0

to co-ordinate and rationalise their operations. Herr Nes is also a member of Norsk Bondelag (Organisation of Norwegian Farmers), which corresponds to the British National Farmers' Union.

The Kvinnherad Kommune provides a relief-man to work on the farm for sixty-eight days per annum, relieving the farmer's family for holidays or in times of sickness.

In Table 10 Olav Nes's farm is compared, as regards production, working hours, costs, income, etc., with all farms in Norway of comparable size, and with some farms regarded as technically advanced. The statistics given are for 1976 (the last year that Herr Nes kept sheep) and were released by the Norwegian Agricultural Economics Institute. In general, the report indicated that Nes Farm was a viable and efficiently run holding. Table 11 gives a detailed breakdown of Olav Nes's income for 1976.

A study of Tjorbeite Farm

Lars S. Mehl's farm (*see* Fig. 92), just east of Rosendal, is part of the original farm estate of Hatteberg, which later, with the other ancient

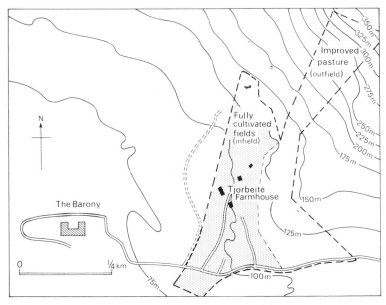

FIG. 92.—*Tjorbeite Farm, Rosendal.*

FIG. 93.—*Lars Mehl's farmhouse, Tjorbeite Farm, Rosendal.*

farms of Mel and Seim, became the estate of the seventeenth-century Barony, which was in turn broken up.

The soil is chiefly composed of glacial boulder clay with detritus from landslides which have occurred on the nearby slopes of Melderskin (1,426 m). With the high rainfall in this region, it is fortunate that the morainic deposits need very little drainage and that heavy machinery can be drawn across it in the worst of weather without becoming stuck, as it would on softer soils. The average pH is about 6.0, slightly acid, as on Nes. The mineral constituents vary from field to field and so, as on Nes Farm, they are dressed with fertilisers appropriate to their needs for the particular crop proposed.

Tjorbeite Farm dates from 1848 when Marcus G. H. Rosenkrone (1823–96) controlled the Barony estate which he had inherited in 1837. This cultured gentleman also devoted himself to establishing a model farm estate and the Tjorbeite farmhouse (*see* Fig. 93) dates from his occupancy, for it was built seven years before his death, in 1889.

In early years this farm was almost certainly a subsistence-type operation, but gradually it moved towards general mixed farming with cattle, sheep, pigs, hens, fruit, including soft fruit, and vegetables, but with some bias towards milk production. Today the farm specialises in producing mutton and wool, but there is some fruit production including red and blackcurrants and strawberries. Herr Mehl also keeps a few horses, a remnant of an indigenous breed.

The total stock of the farm includes 90 Dalasau (Norwegian Leicester × Cheviot) ewes for the production of mutton and wool, with some live-sales. The sheep are injected against liver fluke and internal parasites before lambing, which is from about 15th April to 10th May. The lambs are ear-tagged in aluminium at birth, but no docking, castration or artificial insemination is employed. The sheep receive their first wool clipping and another injection before being sent up to the mountain pastures, which are shared by all farmers who have some part of the original Hatteberg farm or its tenancies. These farmers also cooperate in mustering the sheep back from the hills in mid-September. The practice of taking sheep up the mountain for summer pasture is known as transhumance (*see also* p. 124 and Fig. 6).

After the sheep are brought down they are again clipped, dipped against parasites, and then stalled for the winter months, although they are allowed out to graze for a limited part of each day. From September to November they receive one feed of hay and one of silage per day, but from November to mid-February this is augmented with 1 kg of turnips, and later $\frac{1}{2}$ kg of potatoes. Concentrated grain-feed and minerals are also added, on the basis of researches which have been undertaken in nutrition, in order to ensure, it is hoped, a healthy crop of lambs to follow.

Herr Mehl keeps four breeding mares of the fast-diminishing West

Norwegian or "Fjording" breed, one ideally suited to its environment—lightly built but strong. He keeps them as an enthusiast for the breed, and rears a number of colts for sale.

The farm contains 45 dekar of arable land, and 110 dekar of improved pasture, 15.5 ha in all. While the sheep are on the mountain pasture, the improved pasture is let for grazing cattle and sheep from other farms, and in addition small lambs insufficiently developed to be sent up to the mountain pastures may be kept there.

The arable land is mostly given over to a grass similar in blend to that favoured by Herr Nes. Barely three to four dekar are devoted in equal shares to potatoes and turnips. The turnips (Baugholm variety) yield 6,000–8,000 kg per dekar, and the potatoes (Vestar) 3,000–4,000 kg per dekar. The energy yield of grass is higher when made into silage or hay than when it is grazed (silage yields 450–500 forverd per dekar against 400–450 forverd for grazing, a forverd being the net energy released by 1 kg barley and 14 per cent water).

One-half of the arable land is grazed by the sheep in spring, before they are taken up to the mountain pasture, and again in autumn when they return. While they are away it is let out, as previously noted. The other half is grazed by the horses, and it is in this half that the small plots of potatoes and turnips are rotated with sown grass. The sown grass is also harvested twice a year, in June–July and again in August–September. When part of the arable land is about to be sown with potatoes or turnips it is first ploughed and manure is harrowed into the soil. For turnips there is an additional dressing of 300 kg of calcium per dekar plus some artificial fertiliser. The second year, after a further dressing of manure, the land is sown in grass. On the potato land, no calcium is added in advance, but manure and calcium are added the following year before putting the land back to grass. Herr Mehl does not spray his potatoes or turnips while they are growing. He sows and weeds the plant rows with a horse-drawn machine, but he uses a tractor for ploughing or harrowing. If there is an excess of weeds they are left fallow when the harvest has been completed and the following spring they are ploughed in.

The only timber resource on the farm is a small area of deciduous woodland, birch and alder, which is used for firewood.

Herr Mehl has a tractor which he uses to haul his hay-harvester, livestock boxes, transporters for silage or hay, a front-loader and a roller. He also has horse-drawn carts, a weed-killer, and a machine for cutting V-shaped furrows between rows of potatoes. Additionally he has weed sprays, and equipment for disinfecting and vaccinating the sheep. In his barn he has a grain-drier which operates by blowing air through the grain.

Although the farm is entirely owned by Herr Mehl, he belongs to a "machine ring" where he can rent ploughs, harrows, spray equipment, mowers, trailers, potato-lifters, etc. There is also a "relief ring" for farmers to help one another out with work, and a "research ring" to share the cost of enquiries which will be of benefit to all. The farmers also share a common abattoir at Asane, near Bergen.

The machinery owned by Herr Mehl, along with that made available to him through the "machine ring", does much to alleviate his labour problems, which are particularly severe on a part-time smallholding such as this. Herr Mehl is the chief of the Agricultural Economics Section of the Hordaland Fylke (County) and so the seasonal operations

TABLE 12

Detailed breakdown of Lars S. Mehl's income for 1978

Source	Percentage
Mutton and wool	68
Horses and hens	7
Arable products	4
Government subsidies	18
Other sources (excluding Herr Mehl's full-time income)	3
TOTAL	100

have to be completed at week-ends or in the holidays. Herr Mehl's son also helps with this work in his holidays, and these two undertake all the specialised seasonal tasks as they arise. In winter, however, the evenings draw in, and hired hands tend the animals from Monday to Friday, leaving Herr Mehl and his family to undertake this work as well at week-ends, and in the Christmas and Easter holidays. Thus, as on Herr Nes's farm, most of the labour is provided by the farmer's family. Table 12 gives a detailed breakdown of Herr Mehl's income from the holding.

Future prospects for Norwegian agriculture

In recent years, according to Herr Mehl, there has been a marked change in the Norwegian Government's attitudes, both towards agriculture as a strategic industry, and towards the Norwegian countryside as a valuable natural resource requiring care and conservation. The Government's policies are directed towards national self-sufficiency and the better use of the country's resources.

The Norwegian Parliament (*Storting*) has passed resolutions intended to give agriculture parity of status with comparable sectors of industry. This has stimulated interest in agriculture as an investment. To help promote this expansion reforms will include the provision of regular holidays with full pay, relief services and sickness benefits for farmers as well as farmworkers as a statutory right.

Such reforms will benefit farms like Tjorbeite, enabling the owners to operate them as full-time activities without resort to outside employment, which has become so difficult as to deter people from making their careers on the land.

Mixed Farming in the U.K.

Geographical background

Intensive rotational mixed farming has been the product of a very long evolutionary process in the U.K., where for a long period it has enjoyed government support, and therefore been able to achieve substantial development over a considerable period. It is not operated under the physical strictures experienced by farmers in western Norway, nor has it undergone the disruption and destruction caused by military confrontations fought out in the countryside several times in living memory, as has its Continental counterpart.

Although, therefore, British agriculture was already comparatively efficient, recent economic pressures have, as elsewhere, prompted assimilation of existing farms into larger productive units and into association with one another. The main features of the geographical background of British agriculture are summarised below.

A line joining the estuaries of the Rivers Tees and Exe (i.e. from Middlesbrough to Exeter) divides Britain into the following two distinctive regions.

1. *North-west upland Britain* is characterised by old, hard rocks, a cool, wet climate, and infertile, acid soils rendered thin by glacial erosion. Some of these features are displayed in the Harlech Dome and on Felen Rhyd Fach Farm (*see* pp. 120–5). Such upland farms are characterised by livestock being both their predominant land-use and also their main source of income. The chief livestock are usually sheep and cattle.

2. *South-east lowland Britain* is characterised by young, soft rocks, a rather drier, more extreme climate, giving rise to a more dependable growing season for arable production, and fertile brown forest soils wherein, thanks to lower acidity, the humus is more readily assimilable for crops. Soils are also thicker, having been either supplemented by glacial deposition or, as south of a line from the Severn to the Thames, unaffected by glaciation. Some of these features are demonstrated in the Chelmscote, Broadoak and Linslade Manor Farms, described on pp. 211–15, in that a greater proportion of farm area is given over to crops. Indeed, on some lowland farms the land area is devoted exclusively to crops. Many, though not all, lowland farms also receive a substantial part of their income from arable production, although this is not true of the farm unit cited above.

There are other factors to be borne in mind. First, there is the effect of large, urban centres which may determine land-use patterns in regions within their economic sphere of influence (*see* the theoretical ideas of von Thünen, pp. 3–10, and Ricardo, pp. 77–80, and also the references to the market-gardening and dairying belt around such large cities as London, p. 218 and Fig. 95). Secondly, there are economic trends towards the extension of agricultural associations and the co-operative movement in agriculture, together with growing farm amalgamation stimulated by E.E.C. membership since 1973. These trends are considered in general terms in Chapters XV and XVII, and more specifically in relation to the farm-studies on pp. 120 and 211 (below).

For further background details, the student should consult any good regional textbook of the British Isles.

Chelmscote, Broadoak and Linslade Manor Farms

The farm unit is made up of the original Old Linslade Manor Farm, Broad Oak Farm and Chelmscote Manor Farm. The estate is the property of H. B. and D. M. Leake, but the information embodied here has been provided by the farm manager, Michael Nokes. The farm is situated about 2.5 km north-west of Linslade, Bedfordshire, on the west side of the B488 to Bletchley, Milton Keynes, and consists of about 264 ha of pasture and arable land. The holding is compact, with overall dimensions of about 1.9 km from north to south, and 2.3 km from east to west. Figure 94, however, indicates the practical problems imposed by which are disruption of the farm-lands by roads, railway and canal.

The average size of field throughout the holding is about 5.5 ha. The farmer does not belong to an agricultural co-operative, but he does belong to what is described as a Group Traders' Association, designed to reduce production costs, particularly the purchase of agricultural supplies and agricultural machinery. These economies are achieved through the agencies of A.C.T. (Agricultural Central Trading Ltd), which bulk-buys on behalf of member farmers in a particular district, and also through a machine syndicate of four farmers, one of many such syndicates for which Syndicate Credits acts as guarantor when money is borrowed to buy expensive or specialised machinery, in association with the N.F.U. (National Farmers' Union). The syndicate helps to make it economic to buy machines which are too specialised or too restricted in seasonal use to justify ownership by an individual farm.

There are several distinctive landscapes and soils on the farm. The rather hillier central section, between the Grand Junction Canal and the B488 road, is composed of loam with a high sand content down to about 24 m. This area is liable to constricted drainage as a result of the sand impacting to form a hardpan, and must therefore be subjected occasionally to "subsoiling" in order to shatter the hardpan.

The northern water-meadows of the Ouzel are composed of heavy alluvial silt, while the southern parts of the farm are composed of Oxford Blue Clay. Both these soils have been provided with drains,

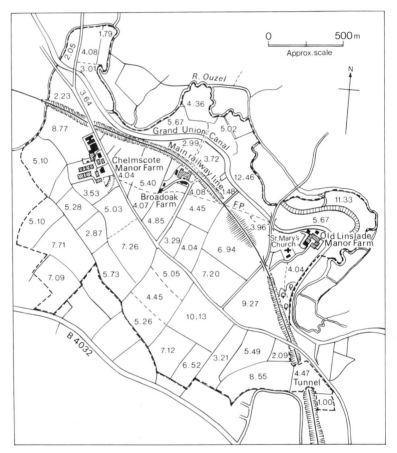

FIG. 94.—*Chelmscote Manor, Broadoak and Old Linslade Manor Farms, near Milton Keynes, Buckinghamshire.* Numbers represent area of fields in hectares.

those along the Ouzel being very old but effective tile drains. Once every eight years or so an implement called the mole is used to bore long horizontal holes through the clay about 460 mm below the surface. These holes and associated radial fractures help to drain and aerate the soil down to the drains, so maximising their efficacy for the next few years.

The climate of the region in which the farm is situated is temperate, with few extremes of temperature:

January mean temperature: 2.2°C
July mean temperature: 16.7°C
Annual temperature range: 14.5°C

The rainfall is moderate and, well distributed, with an annual total of about 635 mm.
The land is utilised as follows:

1. 41.7 ha of winter wheat, varieties Flinor or Atou, which goes for milling, and gives a yield of 5,020 kg per ha;
2. 57.5 ha of spring barley, varieties Luke, Athos, Ambre and Julia, which goes as fodder on the farm, the surplus being marketed; this crop gives a yield of 4,020 kg per hectare;
3. 8.9 ha of forage maize, of I.G. variety, which is used to produce silage for fodder;
4. 28.7 ha of short-term leys (1–2 years) or rotational grass.

The above crops are those involved in the main crop rotation, which includes two years of winter wheat, two years of spring corn, that is, either barley or maize, and two years of grass.

In addition to the short-term leys, or rotational grass, there are other forms of grassland. These include long-term leys of 4–6 years' duration (48.6 ha) and a limited area of permanent pasture. The latter includes about 12 ha of water-meadow near the Ouzel which is liable to flooding and so may not be cropped, although the grass can of course be cut for hay or silage, and small areas of grassland around the three farm-steads, in all 32 ha. The total area of grassland of all kinds is about 109 ha.

Other crops produced on the farm include vegetable crops, e.g. winter onions (Senshya), 2.8 ha, sold for human consumption; lettuce (Avon-defience), 0.8 ha, also for human consumption; lucerne, 9.3 ha; pota-toes, 7.7 ha; and arable silage, 6.9 ha.

There remains the woodland area, which was originally made up of oak, ash, chestnut, alder, beech and cherry. Of these, the beech is now dying of a fungus infestation, but the timber is being salvaged for fur-niture-making. Likewise, the cherry is being used for veneer. It is being replaced piecemeal by some spruce, a quick-growing commercial soft-wood, and also such indigenous species as larch, ash, willow, oak and chestnut. The total woodland area is 5.26 ha.

Finally, buildings and lots occupy the last 12 ha.

As regards the principal enterprise, that of the dairy herd, there are some 420 cattle on the holding. This number is made up of 194 milk cows, 32 in-calf heifers (young females of a year upwards carrying their first calf), 191 "followers" (mostly male calves, or "culls" from the herd, of any age from birth to 18 months old), and finally three bulls, two Ayreshires and a Hereford. The milk output, which is the major source of farm income, varies seasonally from as little as 1,800 litres, to 3,400 litres with summer pasturage at its best. This milk is sold to the Milk Marketing Board, who collect it directly from the farm.

A cow in her prime is usually bred with an Ayrshire bull. Calving is an essential prerequisite for lactation, but also there is a fifty-fifty chance that the progeny will be a cow-calf whose future milk productivity depends upon a good sire, and can be reared for herd replacement. The bull-calves of such a union are, however, of very little value, and are usually sold off after birth for veal production. An aged cow which is due for culling is usually mated with a Hereford bull, or some beef breed. Her last calf is then worth rearing for beef.

Mr. Nokes sells his aged cows, culls, veal calves and "followers" to a meat wholesaler who pays a "dead-weight value price" after the beasts have been slaughtered. Other cattle producers in the area sell their cattle to a wholesaler or dealer, or market their cattle at markets such as those at Aylesbury, Leighton Buzzard, Winslow and Banbury.

Except as regards the water-meadows which are liable to flooding, any consideration of soil types is of only limited relevance to land-use on this largely arable holding. This is because soil dressings are capable of adjusting deficiencies left by the last crop in readiness for the subsequent crop, and because these dressings can also be used to compensate for natural deficiencies in the chemical composition of the soil. Physical limitations, for example variations in water-retentivity, can be adjusted by drainage treatment. Natural soil endowments are therefore far less important than the appropriate treatment to prepare the soil for the intended crop. Winter wheat, for example, is planted on land to which 48 kg per ha of 10–22–22 N.P.K. fertiliser has been applied (10 per cent nitrogen, 22 per cent phosphate and 22 per cent kalium or potash) and where 80–100 units of aqueous ammonia has been injected into the soil. The wheat is drilled at 125–190 kg per ha. Spring barley requires the same quantity of fertiliser, but of 25–10–10 N.P.K., and is planted at the same density as wheat. Finally, grassland receives up to 400 units of aqueous ammonia, cow slurry, and phosphate and potash as required.

The present farm enjoys the amenities provided at the three original farmhouses. For example Linslade Manor includes a dairy with bulk storage for 3,400 litres of milk, a modern milking parlour, cattle-handling facilities, quarters for over 200 cattle, and storage for fodder of all kinds. Chelmscote has four covered yards, additional cattle accommodation and fodder storage, and shelter for machinery, etc. Broadoak has another covered yard and fodder storage.

Machinery, implements and equipment held by the farm include seven tractors, a combine harvester, an assortment of trailers, ploughs, harrows, cultivators, rollers, chain harrows, sprayers, drill-sowers, hay-turners, and a baler. Expensive and less-frequently required machines are held by the machinery syndicate and supplied to the member farmers as needed. These include a hedge-cutting machine, a forage harvester, and a maize precision drill.

The total labour force of eight people comprises one farm manager, two tractor drivers, two herdsmen, one estate and farm management man, one relief milker, and one casual part-time calf-rearer.

The sources of farm income are as follows:

Milk	80%
Wheat, plus a small barley surplus	15%
Miscellaneous vegetables and livestock culls	5%
TOTAL	100%

Chapter XIII

Commercial Horticulture

Introduction

"COMMERCIAL HORTICULTURE" is used here as the collective term for the commercial production of soft fruit and vegetables (market gardening), ornamental and fruit trees (silviculture), and flowers (floriculture). Examples of various kinds of horticultural region in the following paragraphs are all taken from the British context, but later in the chapter there are also studies based in the Netherlands and Norway. Elsewhere, in Chapter III, reference has been made to a leading U.S. horticultural region, California. Such regions spring up wherever favourable natural conditions and/or market requirements justify them.

Production units may display several combinations of the three main activities listed above, or any of a number of local specialisms. Furthermore, such units may occur in very different geographical locations, drawn thence in response to a variety of locational factors.

One region may specialise in the production of a single flower, fruit or vegetable species, e.g. in Britain, tulips around Spalding or roses around Nottingham, raspberries around Blairgowrie, potatoes around Girvan or hops in Kent. This may be the result of cultural or historical influences rather than, or in addition to, genuine geographical advantages enjoyed by that location.

The locational factors which influence the establishment, or ensure the future continuity, of a particular mode of agricultural production in a given area may fall into any of several categories.

First, there are regions of increment, specially favoured in soil and climate, which lend themselves to producing "earlies" in flowers, fruit and/or vegetables. Such regions include the Scilly Isles and the Riviera Coast of south Cornwall, which are able to sell their products at favourable out-of-season prices despite their isolation from their markets, well in advance of less favoured regions of home production (*see* Fig. 95, Type A).

Secondly, there are areas which in addition to these natural endowments also enjoy close proximity to good markets or good communication links with them. Such regions include the Gower Peninsula and adjacent Plain of Gwent, the Vale of Evesham, and parts of the Hampshire Basin, Ayrshire and the South Lancashire Plain (*see* Fig. 95, Type B).

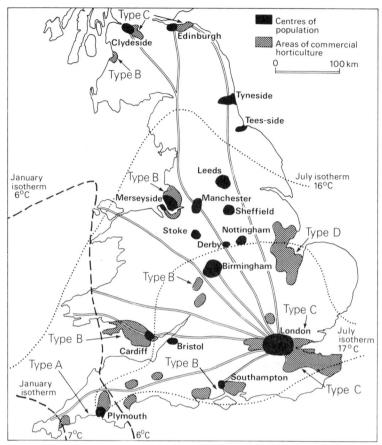

FIG. 95.—*Map to illustrate locational factors behind the main kinds of horticultural region in Britain.*

Thirdly, there are regions, often with little by way of natural advantages, but which produce "perishables" such as soft fruit and vegetables to supply a centrally placed centre of population. Such regions which are peripheral to large settlements are known as their "market-gardening and dairying belts". Today they usually extend about 80 km in radius from the central produce markets (*see* Fig. 95, Type C). At one time such a belt would have extended only about 6–7 km, but expansion has been made possible by faster, more effective means of transport—the farmer is now able to market his own produce each morning by lorry preliminary to his regular daily work. One hour is now sufficient for him to carry his produce 80 km where once a heavily laden peasant could only have taken it 6–7 km. Such physical restraints upon the farmer marketing his own produce preoccupied von Thünen, and the problem is considered in situations all over the world by Chisholm in his *Rural Settlement and Land Use*. London's "market-gardening and dairying belt" includes adjacent areas of Kent, Sussex, Surrey, Middlesex, Oxfordshire, Buckinghamshire, Bedfordshire, Hertfordshire and Suffolk.

Finally, main-crop production areas (*see* Fig. 95, Type D) are regions dependent upon large areas of farm-land, preferably sand or marl, and enjoying a relatively dry, warm summer for the ripening of their crops, rather than the benefit of an early growing season, nearby product markets or good communications. It is significant that Britain's two best main-crop regions are on the hot-summered, drier, eastern side of the country, these being Scotland's east Central Lowlands around Coupar Angus and the Carse of Gowrie, and England's Fenland region. These areas produce the bulk of Britain's main-crop fruit and vegetables, and being unable to supply the remunerative "earlies", depend upon volume of production for their income. The advent of canning, and more recently of frozen food production, has greatly enhanced their income by helping to reduce disastrous gluts which once decimated their profits, but which can now be preserved and sold at some subsequent season when shortage of fresh produce makes prices more favourable.

To illustrate just some of the European diversities within the system, the following examples are cited:

1. silviculture in Haaren, North Brabant, in the Netherlands;
2. floriculture in the Netherlands;
3. greenhouse production on the southern shore of Hardangerfjord, in western Norway;
4. greenhouse production on the L.S.A. Estate at Potton, Bedfordshire, U.K.

Silviculture in Haaren

Of the 5,500 ha given over to silviculture in the Netherlands, about one-third is in North Brabant. De Heer M. F. van Kempen's tree nursery includes three pieces of land: one just outside the northern boundary of the parish of Haaren and on the west side of the Helvoirtse Weg (Way), another just inside the northern boundary on the east side of Helvoirtse Weg, and the third, smallest tract by the Oisterwijke Weg on the western fringe of the town (*see* Fig. 82 in Chapter XII). These total holdings of 25 ha have all been built up in the past ten years from a tiny nucleus of 3 ha. The land was originally "geest", the name given in Dutch to a barren region of sand loam or gravel soils covered in moorland and forest. Since the establishment of the nursery, the land has been cleared and drained by construction of an extensive system of ditches, and in addition a farmhouse and a number of greenhouses and sheds have been built.

The nursery is an entirely independent concern. A wide variety of trees and ornamental shrubs are grown, including conifers such as thuja and cedar, and shrubs such as cotoneaster and euonymus. In addition, heathers such as erica and calluna are produced. Planting takes place in spring, and trees are lifted for sale in the autumn.

The nursery provides employment for a total of twenty-two people, nineteen men and three women, of whom seventeen work full-time, and five part-time. All are general workers capable of undertaking any of the jobs on the holding. Mechanisation includes three tractors, two tree-planting machines, one rotary spade and two machines for potting plants. All the equipment is kept exclusively for work on this holding.

In addition to trees produced for the home market, trees are exported to almost every country in Western Europe by container. Exports account for 40 per cent of the total product, while the home market is accounted for by other nurserymen, 30 per cent of the total, private buyers, 10 per cent, and municipal authorities, 20 per cent.

Floriculture in the Netherlands

In 1977 the total land area given over to flowers and bulbs in the Netherlands was 12,400 ha. This reflects the long-standing and currently growing demand for flowers and bulbs not only from a large home market, but also from all the countries of Western Europe, and even as far afield as North America.

The principal markets in 1977 were West Germany 29 per cent, France 13 per cent, Sweden 11 per cent, the U.S.A. 10 per cent, Italy 8.2 per cent and the U.K. 5.7 per cent, altogether accounting for three-quarters of flower and bulb exports.

The main flower markets are Honselersdyk, Rÿnsburg and Aalsmeer. At Aalsmeer the Auction House is the largest in the world, with an area the size of thirty-six football pitches, and capable, with its five computer-controlled auction rooms, of handling millions of flowers and bulbs during a single session.

A nursery specialising in lily culture is described below. Figure 96 illustrates the lily-cultivating regions of the Netherlands. Of the 12,400 ha given over to flowers and bulbs in 1977, lilies occupied 1,148 ha, i.e. about 9 per cent. However, lily cultivation is of relatively higher value than might be supposed. Although only about 25 per cent of lily output was exported in 1976, compared, for example, with some 60–65 per cent of daffodils and tulips, and in spite of export competition from at least twelve other main flower and bulb varieties, lilies still accounted for 3 per cent of total flower and bulb exports by value.

Peak season for lily production is between April and August, a period

FIG. 96.—*Important areas for lily culture in the Netherlands.* There are only a few growers in North Brabant and Limburg, but far more in the north-west. The Blokkers originated from Akersloot where every year an exhibition is held to show the latest hybrid varieties. This exhibition was first established by de Heer Blokker and two other growers in 1960.

when so many flowers are available that the market price is depressed. Most producers find the high cost of winter-heating their greenhouses deters them from producing lilies out of season. De Heer K. Blokker, whose nursery is described below, produces lilies right through to November.

Lily cultivation at Haaren in North Brabant

Klaas Blokker, helped by his wife Liesbeth and their four sons, is a lily flower and bulb nurseryman in Haaren. The nursery occupies 3 ha in two small fields at the east end of the town (*see* Figs. 82 and 83 in Chapter XII). When he bought it twelve years ago there was only the house, and there were no buildings suitable for their requirements on the sandy, undulating land, which also required draining. Two years later they erected a 1,200 m² greenhouse and a cooling room. (This cooling room, with a capacity of 64 m³, is no longer adequate for their expanding operation.) A year later another 800 m² greenhouse was constructed.

The soil consists of a rich humus overlying sandy material with a loam subsoil. The nursery is not typical of its region and indeed the few nurseries in North Brabant usually concentrate on bulb production rather than flowers.

In January the greenhouse is disinfected and sown with larger bulbs from the crop grown last year in the fields. The varieties of lily in this first sowing are Destiny and Enchantment. Destiny is 600–800 mm tall with lemon blooms spotted in brown, while Enchantment, which grows to the same height, has orange blooms spotted in brown. Both are early to flower, and they are ready for sale between April and June.

In March 1.5 ha of land are ploughed by a contractor, and sown with the small bulbs left from last year's crop. A proportion at least of the bulbs produced will be large enough for sale or for greenhouse production of flowers in the subsequent year. Before planting out, the

[Courtesy Klaas and Liesbeth Blokker

FIG. 97.—*De Heer Blokker selecting blooms in October.*

small bulbs are carefully selected and disinfected, and the tilled land is manured with year-old cow dung. The remaining portion of the land is "rested" by being rented out to a local farmer who may cultivate maize on it. This "rotation" ensures better bulbs in the subsequent season. Land may also be rented from another farmer to grow bulbs on.

In June when the Destiny and Enchantment blooms have been picked from the greenhouse, a second planting is made of *Rubrum speciosum*, pinkish-white blooms spotted with brown, growing on a plant 1 m tall, and Ugida, white with dark pink blooms, and of a similar height. Both are in bloom between August and November (*see* Figs. 97 and 98). Altogether about 120,000 blooms are produced, and yields are about 90 per cent effective in the greenhouse.

The greenhouse has been fitted with an irrigation system which sprinkles the plants early each morning, so that they have all day to dry out and so avoid the risk of bulb-rot which can be a consequence of over-watering.

In summer the field-crop is weeded by another contractor, and sprayed against leaf-eel, virus diseases and mould. There is a rigorous culling of weak or diseased plants. These careful precautions are necessary in view of the fact that a government inspector must approve the entire crop before any bulbs can be offered for sale.

[*Courtesy Klaas and Liesbeth Blokker*

FIG. 98.—*View of greenhouse with lilies coming into bloom in October.*

In November field bulbs are lifted by machine and sorted. The larger ones will be planted in the greenhouse the following season, or some, possibly 10–20 per cent, will be sold. Smaller bulbs are replanted in the field the following year. It may take 2–3 years until the bulbs are large enough for the greenhouse or for sale for flower production. Over the winter the bulbs are all stored in the cooling room, which is kept at a constant temperature of 0°–2°C.

When flowers have been picked in the greenhouse they are sorted into categories: flowers with one, two, three or four, five or six, and seven or more buds are all bound in bundles of ten and packed in plastic cases. Buds alone are discarded. If not required for immediate despatch they are stored overnight in the cooling room. From here they are subsequently transported into Tilburg, 14 km away, where they are loaded on to a refrigerated lorry, again at a temperature of 0°–2°C. The dealer subsequently sells some in the locality or in Belgium, but most are marketed through Aalsmeer (see p. 219), which is 110 km from the nurseries in Haaren, but only 10 km from Schiphol, Amsterdam's international airport, whence they may be flown to destinations all over Western Europe or in the U.S.A. Some are also despatched by refrigerated lorry to destinations in Western Europe. This arrangement with the dealer in Tilburg saves de Heer Blokker the uneconomic journey to market his own flowers in Aalsmeer which he used to undertake regularly at one time.

All the labour, apart from that required for ploughing and spraying which is undertaken by the contractors, is carried out by de Heer Blokker, his wife and sons. During the summer, however, when a lot of additional work may be necessary, young people may be hired as required. Mechanical aids include a two-wheeled tractor used with a trailer, and a digging machine which is only used for lifting bulbs during about one month of the year.

The income of the nursery is derived about 80 per cent from the lily flowers and 20 per cent from bulbs.

Greenhouse Production in Western Norway

In Chapter XII, during the introductory account of the Rosendal District, Western Norway was described as remarkably mild for so high a latitude, and the long-standing cultivation of fruit and vegetables by local farmers was mentioned. However, as subsistence farming has given way to commercial enterprises such as milk production, so too has production of fruit and vegetables for the householders' own use been augmented or supplanted to some degree by specialised producers such as Odd Hus, who operates Fredheim Gartneri (greenhouses) to provide for a much wider urban market than was previously possible. The location of Fredheim Gartneri is shown in Fig. 87 in Chapter XII.

Odd Hus built his first greenhouse, 180 m^2 in area, in 1950, but over the years, with the help of his wife, grown son and daughter-in-law, he has built up a total of six greenhouses with a total area of 2,500 m^2 under glass. The greenhouses are constructed of glass and aluminium, and are fully heated, automatically, by a system fuelled with oil and electricity. Irrigation is also regulated automatically, each species of flower receiving precisely the amount required for that particular species. There are also photo-cells which adjust the supply so that more moisture is made available on sunny days to offset evaporation than on cloudy days. Similarly, an automatic curtain is drawn across the top-glass at night to help reduce heat-loss after sunset, and is withdrawn again at dawn. Fertilisers are automatically dispensed in the water supply.

In 1977 production was given over to 1,500 m² of cucumbers, and 1,000 m² of flowers. In 1978 almost all the capacity was given over to

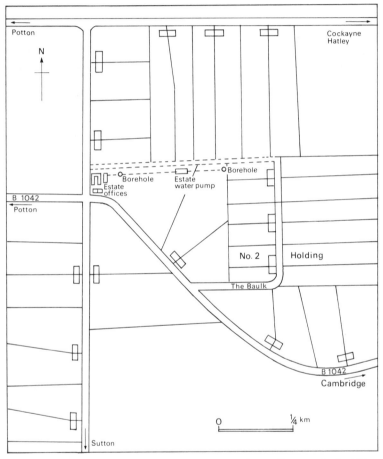

FIG. 99.—*Map of the Land Settlement Association's Potton Estate in Bedfordshire.*

flowers, including more than 100,000 chrysanthemums, 20,000 pelargoniums and azaleas, 9,000 begonias, 3,000 poinsettias, and about 1,000 kg weight of tulip bulbs of various varieties.

The flowers and vegetables are taken by members of the family to market in Bergen (96 km away) or Haugesund (95 km away). The family also provides practically all the labour required, except that a couple of boys or girls may be engaged to help them through the busiest peak season in summer.

Greenhouse Production in Bedfordshire, U.K.

Mr. J. W. Parker is a greenhouseman specialising in salad crops, particularly tomatoes and lettuces. His holding is on the Potton Estate of the Land Settlement Association (L.S.A.), a form of co-operative described subsequently in Chapter XV. A plan of the entire Potton Estate is provided in Fig. 99, while the location of the estate is shown in Fig. 106 in Chapter XV. Mr. Parker's holding, No. 2 on the map in Fig. 99 comprises 2.283 ha, and is roughly 490 × 46 m in extent (*see* Fig. 100).

The L.S.A. Estate is mainly situated on the Lower Greensand ridge which extends diagonally across Bedfordshire into Cambridgeshire, but its eastern edge runs fairly abruptly into stiff Gault Clay. This demarcation occurs on Mr. Parker's holding so that the land marked "Paddock" beyond the "Edward Owen" glasshouse (*see* Fig. 100) is indeed very heavy clay and unsuitable for glass. Most of the land is fairly level, however, so that there is sufficient land on most holdings suitable for glasshouse production.

Only the sandy soil is cultivated on this holding. It is a very "hungry" soil requiring bulk organic manure, i.e. peat or spent mushroom compost, and fertilisers fed in proportions based upon the results of soil analysis undertaken annually after the tomato crop. The soil has a pH of 7.35 according to recent tests. It is also a very free-draining soil because sand particles are larger and so provide greater porosity than would occur in a marl or clay.

The Potton Estate was the first established by the L.S.A. under the original scheme in 1935. At that time much of the land was given over to pigs, poultry and outdoor market gardening. Over the last 12–15 years the swing has been towards glasshouse production of salad crops. When Mr. Parker first took over 10 years ago he had some glass plus outdoor cropping, but as he has built up the glass area so outdoor cropping has declined to negligible proportions.

Mr. Parker keeps 3 stacks of bees and 15 laying hens solely for domestic use, with any surplus going to his family or friends.

Crops include a winter/spring lettuce crop, followed by tomatoes. By heating some houses the planting and harvesting seasons, and thereby the work-load and the marketing season, can be extended. Tomatoes require a very high potash content and a high level of conductivity. By passing an electric current through the soil, and obtaining a high reading, it is possible to establish a high residual content of fertilisers. Lettuce requires the reverse, and therefore when the tomato plants are pulled out in the autumn all excess salts have to be flushed out with large quantities of water prior to planting the lettuce crop.

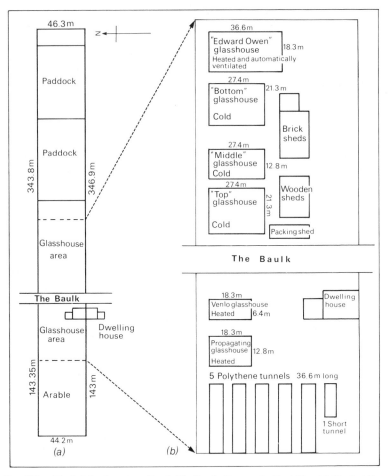

FIG. 100.—*The holding known as No. 2, The Baulk, Potton, Bedfordshire.* (*a*) The holding.
(*b*)The glasshouse area at a larger scale. Area under glass = 2,542 m².

At this stage the soil is usually sterilised with methyl bromide gas by contractors.

In the five polythene tunnels (*see* Fig. 100) the rotation employed is three tunnels of celery followed by lettuce, and two tunnels of lettuce followed by peppers.

Almost all the labour required is provided by Mr. Parker and his wife. Children and friends may help out on occasions. Reciprocal help between neighbours may be called upon in times of emergency, e.g. sickness or when glass has to be replaced as a result of wind damage, and on such occasions all the neighbours rally round.

The six glasshouses (*see* Fig. 100) comprise the following. The "Edward Owen" glasshouse has a steel frame and aluminium glazing bars with sealed glass. It is heated with ducted hot air and automatically ventilated. As the best house, it is used for growing winter lettuce and tomatoes which are planted in late February. The "Top", "Middle", "Bottom" and propagating houses are all steel-framed and fitted with Dutch light glass. Of these, the "Middle" house has ducted hot-air heating, the propagating house is heated by free-blowing hot air, while the "Top" and "Bottom" houses are cold. The small "Venlo" glasshouse is also equipped with free-blowing hot-air heating.

In addition there are several small sheds and the dwelling house. Finally at the west end of the glasshouse area there are the five polythene tunnels used in the production of peppers and celery in rotation with lettuce, as already noted.

Mechanical aids on the holding include a Honda hand-controlled rotivator, which can also be used, when fitted with wheels, to haul a Jack Truck for all the haulage jobs on the holding. A handtruck and a wheelbarrow are also kept. There are two sprayers: a knapsack sprayer and a "Turbair" sprayer. Any heavy cultivation such as ploughing, discing, rolling or cultivating which may be required is undertaken by the L.S.A. on a contractual basis.

All the produce of the holding has to be sold through the L.S.A. under the terms of the contract, and is collected daily from the holding.

Because of the different energy input, and therefore production cost, associated with early tomatoes as opposed to main crop, it is exceedingly difficult to assess the proportions of income derived from the different products. However, the percentages for turnover are tomatoes 70 per cent, lettuces 23 per cent, peppers 4.5 per cent and celery 2.5 per cent.

In addition to furnishing the above information, Mr. Parker was asked his opinion of current economic developments and future prospects for horticulture generally, and for himself in particular. There is currently a problem of overproduction, particularly as regards summer outdoor lettuce and celery. Some apprehension has been expressed over Mediterranean countries, for example Spain and Greece, achieving full membership of the E.E.C., as such countries could supply the British market with out-of-season produce such as early tomatoes so much more cheaply. On the other hand, the rising living standards and production costs in such countries, as well as escalating transport costs, might well offset their advantages, and enable British producers to remain competitive.

Mr. Parker is not unduly alarmed about rising labour costs except in so far as these would be reflected in higher L.S.A. charges for haulage, packing and other services, since he has so few outside labour requirements. He has no plans for changing his present economic cropping programme of salad crops. With the present rising production costs, he has been obliged to increase his turnover during the past two years without any appreciable increase in income. He is not prepared to employ outside labour or to undertake any large investment in new glass, since such expedients, while making possible short-term economies of scale, may, in consequence of inflationary price increases, so add to the operational costs of the holding as to bring about diminishing returns. He therefore prefers to offset such rising costs by increased efficiency.

Chapter XIV

Collective Agriculture: the Kibbutzim of Israel

Introduction

I N this and the following chapter collective and co-operative agriculture are considered. Both provide solutions to what is a growing problem confronting agriculture, that of stepping up production without incurring increased unit production costs, indeed if possible reducing them. One way to do this is to approach optimum size in practical farming unit and farm administration. Collectivisation involves the establishment of huge farm units, where maximum efficiency—optimum size— is achieved in administration, but all too often at the expense of practical efficiency when technology proves inadequate, since the optimum size for farm husbandry tends to be rather smaller. The system is attractive to Communist and Socialist governments concerned to socialise the individual or to achieve rapid economic development. Israel, with her distinctive brand of socialist concern and with difficulties of environment to overcome, is a case in point, and the Kibbutz system is described in detail in this chapter. The co-operative movement, on the other hand, seems to reconcile large administrative units with smaller farm holdings, and to leave the farmer some measure of autonomy. For these reasons it has made ground in the mixed economies of Western Europe, as well as among small peasant proprietors in underdeveloped countries.

Israel: Physical and Historical Background

Physical characteristics of the region

The modern state of Israel was established in 1948 with a population of 879,000 people, 90 per cent Jews, 10 per cent Arabs and Druzes. The country consists of 20,700 km², rather less than the area of Wales. To provide her with security, under the 1948 settlement she was given a demilitarised zone extending far south in the Negev. With this "tail" the country extends 550 km north to south, and varies in width from 15 km to 125 km.

Israel is made up of three physical zones parallel to one another and to the west coast, trending north–south. Most westerly is a coastal plain fringed with low sandy cliffs or sand-dunes. It is in this zone that Kibbutz Gesher Hasiv, described subsequently, is situated a little way south of the Lebanese border (*see* Fig. 101). This fertile Mediterranean zone enjoys the tempering effects of maritime westerly winds, and a winter rainfall a little higher than elsewhere in Israel. It is a region noted for orange groves, orchards and vineyards.

Inland, the central zone is one of hills and plains, with summits averaging about 900 m. This is for the most part limestone country subject to arid summer conditions which make irrigation essential to

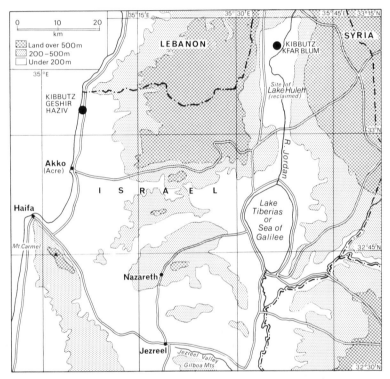

FIG. 101.—*Location of the two Israeli Kibbutzim described in the text.*

sustain its "mixed" crop and livestock economy. Heavy winter rains fill intermittent springs on the hillslopes, but also tend to bring about soil erosion. This hill zone is interrupted in southern Galilee by the Plain of Asdraelon, or Megiddo, which extends south-eastwards in a corridor from the coast between 'Akko (Acre) and the Kishon Defile just north of the Carmel Ridge, inland to Mt. Tabor and the Hills of Nazareth on the north side, as far as Jezreel in the extreme south-east. The hill and plain zone continues south beyond the Carmel Ridge. At the south-east extreme of the Plain of Asdraelon the Jezreel Brook rises, and flows for much of its subsequent course through a valley situated below sea-level, before it joins the even lower-lying Jordan Valley, south of Lake Tiberias (the Sea of Galilee).

The Jordan Rift Valley, El Ghor, third of the north–south trending zones, marks the boundary between Israel and Jordan. For much of its course the Jordan follows a line of tectonic weakness possibly related to those which gave rise to the Red Sea and the East African Rift Valley. The river emerges from several sources on the western, limestone, flanks of Mt. Hermon, 2,794 m high. One of these headwaters, from a re-emergent or Vauclusian spring at 520 m, was built over with a temple to Pan in honour of the Emperor Augustus by King Herod the Great, and this site became known as Panium. This headwater descends 5 km westwards to Dan, another stream flows from a wadi about 460 m above Dan, while yet a third rises on the southern flanks of Hermon. After its descent to Dan, which stands 154 m above sea-level, the river flows through the lands of Kibbutz Kfar Blum, also described subsequently, passing through what was once a marsh and into Lake Huleh, now drained, at about 2 m above sea-level (*see* Fig. 101). This is a distance of some 21 km. Between the site of Lake Huleh and Lake Tiberias (the Sea of Galilee), barely 19 km downstream, the river descends a further 210 m to reach 208 m below sea-level. Before it reaches the northern end of the Dead Sea some 105 km away, it has reached 394 m below sea-level. Thus in the entire journey from the flanks of Mt. Hermon in Syria down to the Dead Sea, a distance of only 150 km, the river has descended 914 m.

The low-lying Jordan Valley has been described as a "torrid and unhealthy cul-de-sac" and indeed, as late as 1943 when Kfar Blum was first settled, the Kibbutz was on the site of malarial swamps. In part of the middle section the river meanders through stretches of drought-resistant forests, which gradually degenerate into bush as the river approaches the salt deserts surrounding the Dead Sea.

The climate of so small a country, as might be expected, is relatively homogeneous. In winter the climate is influenced by weak cyclones which penetrate eastwards through the Mediterranean to Israel from the Atlantic, bringing rain-bearing westerly winds into the country over a brief period of less than two months. Whereas western and higher parts experience a January mean temperature of about 12°C, in the more low-lying Jezreel Valley the average is 16°C. Daytime temperatures in summer may reach 40°C at maximum, and the heat is oppressive. July average temperatures are about 29°C, yet the evenings are often cool and pleasant, thanks to breezes from the north-west which, however, do not bring rain since they emanate from the European summer high-pressure centre.

As regards the winter rainfall, this may be as much as 500–750 mm per annum along the western littoral, but in southern and eastern areas, e.g. the Jordan Valley south of Lake Tiberias, it averages only 250–

500 mm. Although briefly the little watercourses in the dry limestone hills may be full of water, some 92 per cent will be lost to farming as a result of rock porosity, run-off and evaporation.

Past human development in the region

Although only a proportion of modern Israelis are orthodox Jews, the aspirations which led to the establishment of the modern state of Israel and the idealism of the Kibbutzniks are not wholly explicable without reference to the traditions and beliefs from which they have derived their purpose. The original Jewish stock probably sprang from southern Mesopotamia, whence they appear to have undertaken a succession of migrations around an ancient and well-watered trade route known as the "Fertile Crescent" to enter Canaan, as the region was then called, from the north-east in about 2000 B.C. After a subsequent period of sojourn in Egypt, they re-entered what they regarded as their "Promised Land" under Moses in about 1290 B.C., and eventually took possession of it by force. The Kingdom of Israel reached its zenith in the ninth century B.C. under Kings David and Solomon. Thereafter the by then disunited nation came under domination and cultural influence from Assyrians, Chaldeans, Persians, Greeks and Romans, but never did this little nation wholly lose its sense of divine mission and identity. In about A.D. 4 a man was born among them destined to wield a world-wide influence throughout the subsequent two millennia, Jesus of Nazareth. Despite his claim that he stood for the fulfilment of all their religious aspirations, however, he was rejected by his own people. Scarcely a generation passed before many Jews were obliged to leave their homeland as a result of Roman persecution. In the subsequent period of a little short of 1,900 years, the Jews were dispersed widely throughout the world. Nevertheless, they retained their strong religious and cultural identity, even when these placed them counter to the people amongst whom they settled. Persecuted, often without legal rights of ownership, they have engaged in commercial enterprises which could benefit from international contacts made easy by their dispersal, and avoided involvement in more sedentary occupations. Their success in business has aroused jealousy, while their international collaborations have aroused suspicions of conspiracy, a distrust heightened by their refusal to lose their Jewish identity. The climax of anti-semitism must surely have been reached when the Nazis, in pursuit of a solution to the so-called "Jewish Problem", killed 6 million Jews between 1939 and 1945. Such acts of atrocity served to rekindle in many European Jews the age-long resolution to return to their "Promised Land"—the spirit of Zionism was now intensified.

Thus in 1948 representatives of European Jewry who had survived the Nazi terror, along with British and American Jews disenchanted with life in lands so alien to their culture, plus minorities from various parts of the Middle East and North Africa, began to return "home" to join some already living amongst the Arabs in the Palestinian Mandate. In doing so they received earnest international support, but aroused the hostility of the Arab World, since it entailed displacement of Arab farmers, and the emergence of a new non-Arab power in their midst when the Jews re-established themselves in their "Promised Land".

The leaders of the new Israeli State have been confronted with the problem of trying to integrate into one nation Jews from so many countries and social backgrounds. However, the unifying effects of religious belief for some, together with common Zionist aspirations, have helped most if not all the immigrants to settle and work together. In this book we are concerned particularly with the agricultural development of Israel.

Agricultural Developments in Israel

Many Jews, both those who infiltrated into Palestine before 1948 and those who joined the new Israel, had been influenced by:

1. the Zionist Movement, which nurtured their ambitions to re-establish a Jewish state in the ancestral homeland;
2. the Jewish Law and Prophets (the Old Testament) which so long ago had been based upon notions of social justice and equality in the sight of God;
3. Utopian socialism, which stood for the setting up of a new kind of society holding property in common, and in which the individual works for the good of all.

As early as 1882 a Russian Zionist Youth organisation, the "Bilu" Movement, encouraged many young Jewish idealists to emigrate to Palestine. The earlier ones failed to sustain their ideals in the face of harsh realities, but later, in the 1900s, settlers whose Zionism was tempered with socialist convictions managed to establish themselves successfully on land purchased from Arab landowners by the Jewish National Fund (J.N.F.). The fund was set up in 1901 for the purpose of buying up land in Palestine for Jewish settlement, and the founders intended that the land should be held in perpetuity on behalf of the entire Jewish people and not made available for repurchase by individuals.

Forms of agricultural organisation

Over the years, Jewish settlers included socialist Zionists from the U.S.A. disillusioned by the collapse of the "American Dream" as a result of the economic depression after 1929, and from Europe following the beginnings of Nazi persecution soon after 1933. These settlers adopted various forms of farm organisation, including the Moshav Shitufi, the Moshav 'Ovdim and the Kibbutzim, which emerged alongside privately owned agricultural communities or "Moshava"—the characteristic pattern of agricultural holding among the citrus producers of the coastal lands.

The Moshav Shitufi are collective smallholders' villages, in which the farmers employ no hired labour, purchase their requisites and sell their produce through a co-operative, and share the profits collectively. The Moshav 'Ovdim are co-operative smallholders' villages, similar to the above except that individual members retain their own profits. In both these kinds of organisation, the family is retained as the basic social unit, and in this they differ from the Kibbutzim. Further information about Israeli co-operative agriculture may be obtained from the Plunkett Foundation for Co-operative Studies (*see* Bibliography, other sources of information).

The Moshav Shitufi and Moshav 'Ovdim, however, fall short of the collective ideal epitomised by the Kibbutz. It is the nationalist zeal, socialist idealism and powerful motivation of the Kibbutznik which has made possible the pioneering of agricultural projects in the most inhospitable of conditions, facilitating the reclamation of swamps and deserts, and the achievement of a high level of self-sufficiency by the Israelis. Irrigation, by diversion of water from the only perennial rivers available, the Jordan, Yarkon and Kishon, pumping up of subterranean water reserves from beneath the coastal plains, construction of reservoirs, and desalination schemes, has increased the land area available to agriculture. Productivity has been enhanced by employment of soil conservation techniques such as improved rotations, better fertilisers, terracing and reafforestation of slopes, and the adoption of mechanised techniques of dry-farming. More recent crop introductions have included cotton in the Plain of Sharon, dates and bananas in the Jordan Valley, and wheat, barley and fodder crops in the Negev.

Growth of agriculture and of Kibbutz Movement

It should be noted that the generally accepted unit of land measurement in Israel is the dunam, which is closely comparable in size to the Maltese tomna (*see* p. 179) and the Norwegian dekar (*see* p. 197). Like these, the dunam reflects the small units of land operated as farms in the past.

$$1 \text{ dunam} = 0.1012 \text{ ha}$$
$$1 \text{ ha} = 9.88 \text{ dunams}$$

In 1948, when the population of Israel stood at 879,000, the irrigated area was 290,000 dunams (29,340 ha), and the total arable area was only 1,650,000 dunams (166,936 ha). At that time 18 per cent of the population was engaged in agriculture. In 1959 the irrigated area had been extended to 1,250,000 dunams (126,467 ha), and the total arable

area had been increased to 4,000,000 dunams (404,694 ha).
population of 2.18 million, including 1.8 million Jews, 18 per
still engaged in agriculture. In 1974 the total population of
estimated at about 2.75 million, of which, 3.8 per cent or about 100,000
were Kibbutzniks, compared with 80,000 in 1961.

The Kibbutzim: in Ideology and Practice

Yosef Criden and Saadia Gelb together contributed to the remarkable
appraisal of the Kibbutz, *The Kibbutz Experience—Dialogue in Kfar
Blum*, and their words are quoted here.

Concerning Kibbutz ideology:

> There are ... two fundamental concepts ... which are basic to Kibbutz ideology.
>
> First of all, there is the idea that you are working to support not only yourself or
> your immediate family but also your haverim (comrades). This is in accordance with
> the classic socialist principle—"From each according to his ability, to each according
> to his need". No one exploits anybody else. All live together as one family, working
> for the common good.
>
> Secondly, we are all dedicated to the Zionist Cause—we are working together to
> rebuild the Jewish homeland.
>
> The idea of the Kibbutz was not just to achieve co-existence between Zionism ... and
> ... Socialism, but a fusion of the two. The Kibbutz was part of the workers' movement,
> the aims—to take Jewish intellectuals and turn them into workers .... to evolve a cadre
> of pioneers devoted to the service of the nation ... to turn out a new better type of
> human being.

What these ideals entail in practice:

> In Israel the various Kibbutzim, from the very outset, tended to band together into move-
> ments, culminating in a federation of Kibbutz Movements. As an integral part of the
> labour movement, the predominant sector of Israeli society, the Kibbutz was not only
> not isolated from society—but played a major role in the national and social revival
> of Palestine.
>
> The Kibbutz is essentially a group with common economic, social and political ideo-
> logies. It was founded on the idea of Jewish farmers working nationally owned land
> through collective ownership by the whole group of all the means of production—
> soil, equipment, livestock and so forth. The commune itself controls purchasing, pro-
> duction, marketing, consumption, including housing, the rearing and education of the
> children, cultural, recreational and all other services. No wages are paid; every member
> is expected to work according to his ability, and has all his basic needs met in return.
> There is complete equality in both rights and obligations. A necessary corollary to
> the requirement of self-labour was the rejection of all hired labour, a principle accepted
> by most Kibbutzim. According to the model rules for Kibbutzim, drawn up as the
> Kibbutz movement grew, a member leaving the Kibbutz is not entitled to "withdraw
> any private belongings or monies which he may have turned over to the Kibbutz when
> accepted as a member", unless specifically provided for ... In theory at least, the Kib-
> butz forms a single estate, a community, a single large household which is responsible
> for the needs of every member. The individual family remains unaffected economically
> by illness, invalidism or the death of the head of the family. The leadership is demo-
> cratically elected and enjoys no special privileges; management of the various branches
> of the Kibbutz economy rotates, and all decisions on the economic, social and cultural
> life of the Kibbutz are made by the members themselves.

Even such explicit rules of practice as these have required modification in the face of social and economic change, however:

> You organised a group of dedicated people who would live together, go through struggles together, make mistakes, and evolve a new life-style of their own. That is why there [is] really no blueprint for the Kibbutz. Beyond certain clear-cut basic concepts the details developed differently from Kibbutz to Kibbutz. Naturally, over the years, some of the original naïve idealistic aspirations have undergone considerable change ...

The Kibbutz was originally conceived as an agricultural community, but:

> By the early 1930s industry had been introduced into some of the Kibbutzim ... theoretically to take up the slack during the less busy seasons on the farm. But no modern factory with a production schedule can rely on seasonal unemployment on a farm for its labour force—it must work on a year-round schedule. Furthermore, its labour force becomes increasingly more sophisticated and specialised as Kibbutz industry becomes more complex, since it requires skills the average Kibbutznik does not have. Industry could not remain a part-time objective ...

Industry on the Kibbutz has improved agricultural technology and so helped to step up agricultural production. It provides work for older and infirm people, and women unfit for the more strenuous employment on the land. It has also changed attitudes to higher education:

> Whereas most Kibbutzim had been reluctant to let their youngsters go to college for fear they would become "over-educated" for life in a comparatively simple agricultural economy ... they now want their children to take advantage of the opportunities for higher education so as to bring added skills back to the Kibbutz.

"An unanticipated benefit of Kibbutz industry has been the success of workers' participation in management programmes." The Kibbutz has also reversed the trend of the Industrial Revolution which "drained the villages of their labour force ... the Kibbutz brought industry back to the village" and so has served as a "deterrent to the flow of younger people out".

Such changes have not been simply economic. The austerities of early Kibbutz life are giving way to a more liberal and congenial life-style, for example:

> One of the founders of the Kibbutz movement ... said that your room should consist of four walls, a floor, a ceiling and a bed on which to rest your head ... Everything else, you were supposed to do as a community. Today ... your living quarters ... will include a kitchenette ... it is no longer considered a heinous crime to want to spend your evening alone or with ... family or friends.

The younger generation is even asking for opportunities to travel abroad.

Contributions of the Kibbutzim to Israeli society

> Today there are about 250 Kibbutzim in Israel. Their total population is over 100,000, including members and their children, parents of members, groups in training, and

other categories of permanent residents. The largest of the Kibbutzim have populations of more than 2,000, some as few as 100. In the majority of Kibbutzim the population is between 200 and 700.

With only 3.8 per cent, perhaps a little less, of the total population of Israel, the Kibbutzim produce half its food ... 7 per cent of Israel's industrial product .... with close to 250 enterprises in over 160 Kibbutzim. Of this, 25 per cent is exported, in 1972 Kibbutz industry brought in over 45 m. dollars.

It is obvious that there would have been no Israel today without the Kibbutzim, which reclaimed and rehabilitated the land, developed a broad spectrum of agricultural products, established a Jewish presence up to the very borders of what was to become Israel, and provided leadership for the army even while it fed the population at large ... It can still play a major role in the absorption of new immigrants and the development of new industries.

From the preface by Mordecai S. Chertoff to
The Kibbutz Experience

Kibbutz Kfar Blum

Kfar Blum was named after Leon Blum, a French premier imprisoned by the Nazis at the time the Kibbutz was established in 1943. It is situated in northern Galilee, in the Upper Jordan Valley (*see* Fig. 101). A plan of the Kibbutz is given in Fig. 102, while Fig. 103 shows how the Kibbutz looked in its early days, in 1947. Yosef Criden reminisces:

The River Jordan flows right through out Kibbutz. On one side we have the Golan Heights from where the Syrians used to look down upon us ... On the other side, you can see the hills of Lebanon. The house I live in sits on ground which used to be under water ten months out of the year, because this was once the northern end of the Huleh Swamp. There was a period when we had 100 per cent malaria.

Now, however, the land is drained and the mosquitoes are no more. Figure 104 shows some of the housing in Kibbutz Kfar Blum today.

The subtropical climate is characterised by hot summer days—40°C is not uncommon at midday in July—but with cool, pleasant evenings. With midday January temperatures of about 20°C, the winter is relatively cool, and evening temperatures may fall to about freezing-point. The rainfall has averaged 514 mm per annum over the thirty-four-year period, 1944–78. However, most of this rainfall is concentrated in the winter quarter of the year, as shown by averages over the same long period:

Winter qtr.	Spring qtr.	Summer qtr.	Autumn qtr.
(Dec.–Feb.)	(Mar.–May)	(June–Aug.)	(Sept.–Nov.)
326 mm	114 mm	0·7 mm	73 mm

Not only is this rainfall unevenly distributed, but it is very unreliable from one year to another. Thus while the average rainfall of the winter quarter from 1944 to 1978 was 326 mm, it has ranged from as little as 176 mm to as high 603 mm over the same period. These figures, recorded at the Kfar Blum weather station, serve to demonstrate the water problem which confronts agriculturists in Israel.

Population structure and sources of income

The 1943 settlers consisted of about one hundred young people who had already undergone five years of preparatory training at a centre on the coast. They included Lithuanians, Latvians and Estonians from the Baltic area, British, Americans and Canadians. Subsequently, the Kibbutz expanded most rapidly in the years following the establishment of Israel, so that in 1960, of a total of 657 people on the Kibbutz, 265 were members or candidates for membership, with 22 parents of

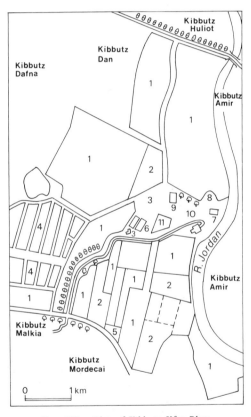

FIG. 102.—*Plan of Kibbutz Kfar Blum.*

Key

1. Cotton
2. Orchards
3. Poultry
4. Fishponds
5. Pecan nuts
6. Heated glasshouses (roses)

7. Regional High School
8. Swimming pool
9. Guest House
10. Living area and services
11. Cattle sheds

[*Courtesy Archivist, Kibbutz Kfar Blum*

FIG. 103.—*Kibbutz Kfar Blum in 1947.*

[*Courtesy Archivist, Kibbutz Kfar Blum*

FIG. 104.—*Housing in Kibbutz Kfar Blum today.*

members, 290 children, and 80 others. Since that time, the Kibbutz population has tended to age—the original pioneers are now middle-aged, and there are fewer children in the settlement. Thus, in October 1978, of a total population of 700, there were 411 members or candidates for membership (387 and 24 respectively), 12 parents of members, 186 children, and 25 students—mostly American students on a ten-month educational programme. In addition there were 66 "others", mainly temporary groups, individuals or volunteers. At the same date the 387 full members were made up of 196 men and 191 women, and of 334 married people (167 couples) and 53 single people. A check on the ages of members undertaken in March 1978 revealed that, of the 373 members at that time, 15 were over 65 years of age.

The economy of the Kibbutz is primarily based on agriculture, which is described in full below, but the Kibbutz has supplementary sources of income as follows.

1. People who hold employment outside the Kibbutz but whose salaries are paid into its treasury include those employed in regional and co-operative industries and plants—the slaughter-house, cold-storage plant and laboratory, bakery, fruit-packing plant, and cotton gin—and also those who teach locally, are employed in government offices or the Galilee Regional Council offices, or serve in the National Kibbutzim Organisation. These total forty-six.

2 Education and child care: the Children's House, where the very young Kibbutz children live, and the Regional Elementary and High Schools employ in all sixty-four people.

3. "Electro-mechanical Products", Kfar Blum, are manufacturers of several kinds of illuminated synoptic control and display panels individually designed to meet the needs of particular industries. In all this plant employs another twenty-four people.

4. Workshops include the dehydration plant worked in conjunction with several other Kibbutzim to process alfalfa, machine shops, a garage, carpenters' and electricians' workshops, and a smithy. These establishments provide work for another eighteen workers.

5. The Guest House, which has excellent facilities including two tennis courts and an Olympic-size bathing pool, can accommodate guests in thirty-eight double rooms and provides work for fifteen women.

6. Internal services of the Kibbutz include the kitchens, laundry, shoe-repairs, sanitation, landscaping, clothing shop, supermarket, athletics facilities, book-keeping office, library, archives, craft-hobby shop, nature museum, a construction department and all facilities required for a self-sufficient community. Between them such facilities require a staff of 148.

7. Occupations related to agriculture, in addition to some of the workshops already listed, employ a further forty-three persons.

8. Lastly there are soldiers of the regular army, members on long vocation—three months to a year—sick and disabled people, etc., making up the balance of approximately twenty-nine people. Of course, these figures are very approximate, as they vary from day to day.

The farm at Kfar Blum

Today the farm remains the major economic activity carried on in the community. The productive farm area of 5,000 dunams (506 ha) lies in close proximity to the property of the Kibbutz. Again, quoting Yosef Criden:

> Ours is a setting which people sometimes describe as a country club. There are spacious lawns, beautiful trees, a magnificent swimming pool, and a wooden-bench sauna. There is a fine well-equipped High School, which serves not only Kfar Blum but the entire surrounding area, and a nature study house. There are comfortable living quarters, a handsome dining hall, and farm buildings of every description. There is a factory, there are sidewalks, there are nooks where people can sit—and places where you can go for a walk in the evening. There are tennis courts and fish ponds, and orchards and fields stretching out in every direction.

Field crops occupy an area of 4,000 dunams (405 ha). Chief among these crops is cotton, to which 2,650 dunams (268 ha) are devoted. This crop provides nearly one-third of farm income. The soil is dressed every year with ammonium sulphate, and the crop requires constant irrigation with sprinklers. The water is mostly obtained from the Jordan in late spring when melt-water from Mt. Hermon replenishes the river. The cotton is marketed through the National Cotton Marketing Board and 50 per cent is exported. The remainder is sold to local textile mills, which also export their finished goods.

The remaining area of 1,350 dunams (137 ha) is mainly devoted to fodder crops, alfalfa and maize, for cattle and poultry, some flowers and vegetables, etc.

Orchard crops occupy 500 dunams (51 ha) of land and are planted with mature apple and pear trees. The orchard area is fertilised with ammonium sulphate and a urea spray, and a drip system of irrigation is employed in the dry season. The average crop yield is about 4,000 kg per dunam (about 40,000 kg per ha). The 1977–8 crop was somewhat lower than average, with 711,000 kg of Grand Alexander, 406,000 kg of Jonathan, 356,000 kg of Golden Delicious for apples; and 51,000 kg of Peckham, 102,000 kg of Gvar Am (a local variety), 122,000 kg of Orleans, and 152,000 kg of Delicious for pears. The fruit is marketed by Tnuva, Hebrew for yield, Central Marketing Co-operative of Histadrut, General Federation of Labour. Orchard crops contribute another quarter of farm income.

The three poultry houses contain 100,000 birds of the local "Anak" breed (White Rock × Cornish). Four batches of birds are reared in a year, making a total of 400,000 birds in 1977–8, and representing 660,000 kg of chicken meat. It was planned to step up production to produce 650,000 birds in 5 batches of 130,000 in 1978–9. Ninety per cent of the birds are sent to the local slaughterhouse owned jointly by the local Kibbutzim and the Regional Council, while the other 10 per cent are marketed live through Tnuva. This production is worth nearly a third of the farm income.

As regards the cattle, the project was originally developed to provide manure for the land, before synthetic fertilisers became so cheap and readily available. Israelis are not big milk drinkers but of late demand for milk has risen to supply an expanding cheese-making industry. Because of the high level of investment involved in operating the dairy farm, and the cost of capital equipment—modern cow-barns, milking machines, etc.—as well as the intensive labour requirements, this aspect of the farm economy is currently under review. There is a herd of 210 Friesian cows and one breeding bull, and approximately 200 calves, half of them male, half female, are produced in a given year. The male calves are sold for meat at four days old, and the female calves retained for subsequent herd replacement. The milk yield in 1977–8 was 1,630,000 litres, which works out at 8,400 litres for every cow currently in milk. The milk and calf meat or veal is marketed through Tnuva, the calves being slaughtered at the local slaughterhouse. In addition, the Kibbutz benefits from 2,000 m³ of organic manure. The dairy cattle provide a further 10 per cent of farm income.

Finally, 500 dunams (51 ha) are given over to fish-breeding ponds. In the 60 dunams (6 ha) of carp pond yields of 1,500 kg per dunam, have been reached. In 1977–8, 100,000 kg of carp was sent to market and a further 30,000 kg went to industry. After young fry of St. Peter and Grass Carp have reached 300 g they are transferred to large ponds of 375 dunams (38 ha) in extent. These ponds are supplied with oxygen and the fish are fed with concentrated food pellets. Output in 1977–8 included 41,000 kg of St. Peter and 30,000 kg of Grass Carp marketed through Tnuva.

Since land furnishes most of Kibbutz employment, the farm is comparatively labour intensive. Field crops are managed by ten skilled full-time male workers, including those who specialise in irrigation, tractor-work, pest control, etc., aided by three part-time female workers. The orchards require twelve full-time workers, eight male and four female, of which eight are skilled, and in addition six part-time helpers. During the picking season as many as sixty people may be employed in the orchard each day. The poultry require four full-time workers, of which three are skilled, and one part-time helper. All the full-timers are male. The cattle management and dairy provide employment for seven full-

time staff, of whom five are skilled, and three part-t
Finally the fish-ponds require three full-time a'
workers, all male. Three of these fish-pond workers

Despite this relatively intensive use of labour, the level of
tion is still quite high, since although the aim is to give everyon..
poseful employment on the land for at least some period of their work
ing life for its therapeutic value, it is not intended that it should involve
unnecessarily arduous toil, nor that productivity should suffer for want
of essential equipment. There are thirty-five tractors, one lorry, three
cotton-pickers, a selection of sprayers and other equipment. There is
also a selection of small cars and multi-purpose vehicles, all of which
are communally owned, for use on the farm or on business, or for
members' personal use.

The percentages of farm income contributed by the various sectors
of production are estimated as cotton 30 per cent, orchards 25 per cent,
poultry 30 per cent, dairy produce 10 per cent and fish 5 per cent.

In assessing the success or failure of the Kibbutz system of agriculture
there is a danger that this is seen in terms of output value in excess
of input costs, i.e. profit. This is, however, contrary to Kibbutz thinking
as Yosef Criden explains:

> The efficiency of a Kibbutz must be analysed in the light of many different factors.
> For instance, our 1,250 acres [506 ha] could probably be managed more efficiently
> by one ... farmer with a large family and perhaps a few hired labourers, than on the
> basis of a Kibbutz with several hundred members. From that point of view, the Kibbutz
> is probably not run on an economical basis. On the other hand, if we consider that
> we have been able to make our 1,250 acres support over 600 (actually 700) people in
> relative comfort, we can say that, in fact, the Kibbutz is operating in a very efficient
> manner.

To identify those features common to most Kibbutzim, Kfar Blum
may be compared with another similar establishment.

Kibbutz Gesher Haziv

Gesher Haziv is situated near the coast of Israel, about 1 km from the
sea and 5 km from the Lebanese border (*see* Fig. 101). As a consequence
it is within range of Lebanese rockets and artillery. A plan of the Kibbutz is shown in Fig. 105. At the moment there are 465 people living
in the Kibbutz. Gesher Haziv was founded in 1949. From the very beginning the community there rejected one of the basic rules of the movement, by deciding to have their children sleeping in the parental home
rather than in a communal children's home, as was then the general
rule. Although at the time condemned this has now become standard
practice.

Population structure and social organisation
Members and candidates total 255, 128 male and 127 female, and children, 74 male and 48 female, make another 122. There are 19 parents

Fig. 105.—*Plan of Kibbutz Gesher Haziv.*

Key

1. Turkey houses: breeding section
2. Turkey house: controlled environment, dark room
3. Cotton and potatoes
4. Bananas
5. Avocadoes
6. Plastic factory
7. Citrus fruit: oranges, pomelos, tangerines, grapefruit
8. Regional High School
9. Guest House
10. Living area and parks
11. Turkey houses: fattening pens
12. Brooder house: capacity 20,000 poultry
13. Machine shop, garages and related farm buildings
14. Winter crops

of members, and almost equal numbers of both sexes of volunteers, 69 in all, living there temporarily, making a total of 465 people, amongst whom there are 89 married couples. The founder members, 50 in all, are all between 48 and 65 years of age, and the Kibbutz will soon have to face the problem that an increasing proportion of its population is old, a situation already affecting some older Kibbutzim. There are, on the other hand, 66 members who were born or grew up on the Kibbutz, and 142 residents between the ages of 19 and 45 years.

The Kibbutz is governed by a Praesidium made up of the General Secretary, Farm Manager, Work Manager, Treasurer, Chairman of the "Personal Problems" Committee, and two members at large. Every facet of Kibbutz life—employment, culture, health, education, finance, buildings and landscape, sports, etc.—is regulated by a committee elected by the General Assembly, usually for a one-year term. All decisions are finalised at a General Meeting attended by all the members once a fortnight. Anyone joining a Kibbutz is first a "visitor" for one month. If approved by the Absorption Committee, the person stays for a further eleven months as a candidate. After this, the decision to accept or otherwise is made by the General Meeting, and a two-thirds majority is required in the secret ballot. Not merely joining candidates, but husbands, wives, even children of members born on the Kibbutz, must be elected in this way.

The work-force comprises 90 per cent of the adult population, or approximately 290 people. Children begin by working six hours per week in various sectors of the Kibbutz at 14 years of age. About 10 per cent undertake further studies, long or short courses directly or indirectly connected with work on the Kibbutz, after completing their twelve years of compulsory education.

Sources of Kibbutz employment

Agriculture provides full-time employment for forty-two people, including four women. This includes work in fields, orchards, and livestock management. Volunteers are usually employed in land work irrespective of sex. In addition another fifteen people work in occupations related to agriculture, making a total, including the volunteer workers, of 125 directly or indirectly involved with the Kibbutz farm. Agriculture is also the main source of Kibbutz income, and so in order to be Farm Manager a member must study for four years at university, since today in Israel agriculture requires personnel of a very high calibre.

The remaining 165 or so workers are either in industry, Kibbutz services or outside employment.

Industries include a factory producing plastic goods such as pallets for shipping canned goods, a child's high-cum-rocking chair, and components for the defence industry. There is a guest house which utilises the home-grown foodstuffs, and offers seventy-six rooms for the use of guests, who include, in addition to tourists, local people seeking a haven of rest.

Kibbutz services range from catering, laundry, child-care work and education or nursing, to farm or factory management, maintenance work (staff include electricians, plumbers, repair-men and drivers) and administration.

Finally there are members employed outside the Kibbutz whose wages are paid directly into the Kibbutz treasury. The Kibbutz is an active participant and shareholder in a co-operative conglomerate whose main concerns are with the processing and marketing of the agricultural produce of the region. The conglomerate operates a feed-mill which utilises various field crops to produce a range of specialised animal feedstuffs, a cotton-gin and refinery for the production of seed-oil and protein, a slaughterhouse where poultry are processed to the point where only water remains and even this is recycled, a fruit-packing plant, a fruit-canning factory making a range of canned juices, marmalade, etc., a meat-canning factory producing processed meats, frozen chickens, etc., and finally a computer service for all the Kibbutzim and factories within the conglomerate.

The farm at Gesher Haziv

In this coastal region winters are comparatively mild, with January daily maximum temperatures as high as 17°–20°C, but nightly temperatures falling to 7°C. Summer temperatures may actually rise to about 37°C, although the average is rather lower. Prevailing winds are from the south-west and north-west, and mainly light. In the autumn "Chamsin" winds come from the desert—in Arabic this word means "fifty", because the Arabs once said that these hot, dry winds blew for fifty days. The reafforestation which has taken place in recent years has helped to mitigate the enervating effects of these winds. The average rainfall is 650 mm per annum, all falling between November and March, while the remainder of the year is almost wholly without precipitation. The relative humidity is as high in summer, however, as in winter.

The Kibbutz is situated 30 m above sea-level. The soils are composed of alluvial material, mainly limestone or sandstone detritus eroded from the hills of Galilee, producing a light, well-drained soil. When necessary this natural drainage is supplemented by the construction of concrete drains. There are 3,460 dunams (350 ha) of land on the holding.

Cropping system. Table 13 shows the area under each crop, and the yields that have been achieved.

1. *Irrigated crops* include cotton and potatoes. The land available for cotton is restricted by the availability of water for irrigation. No rotation is practised. Ammonium sulphate and potassium chloride are

TABLE 13

Crops and yields on Kibbutz Gesher Haziv

Crop	Area cultivated dunams (ha)	Yield, kg/dunam (kg/ha)
Cotton "Akala", cash crop for home and export	850 (86)	400–500 (4,000–5,000)
Potatoes "Patronas" and "Desirée", cash crop for home market only	150 (15)	4,600–6,100 (45,000–60,000)
Wheat Cash crop for home market	[variable]	300–400 (3,000–4,000)
Bananas Large and small "Cavendish" (Australian introductions), cash crop for home market	400 (40.5)	—
Avocadoes	540 (55) [351 (35.5) fruit-bearing]	1,000–2,000 (10,000–20,000) [mature trees only]
Grapefruit	95 (9.6)	6,600–9,100 (65,000–90,000)
Valencia oranges	65 (6.6)	4,100–4,600 (40,000–45,000)
Washington navel oranges	25 (2.5)	4,600–5,600 (45,000–55,000)
Jaffa oranges	10 (1)	4,600–5,600 (45,000–55,000)
Tangerines	65 (6.6)	3,000 (30,000)
Pomelos (shaddock)	20 (2)	[yield unknown, not yet sufficiently mature to bear fruit]

applied during the growing season. The cotton crop is planted in March. Defoliation is done by the spraying planes, and the crop is picked in October. The crop is marketed through a co-operative called MILUOT, owned by the farming settlements along the coast between the Lebanese border and Haifa Bay. The cotton is processed at MILUOT either for local use or for export. By-products are cotton-seed oil, and synthetic protein for animal and human consumption.

The potato crop is rotated on a four-year plan. The land is dressed with ammonium sulphate and potassium chloride. Planting is in February, and irrigation by sprinkler at the rate of 500 m³ per dunam (4,900 m³ per ha) supplements whatever light rain may fall after February. The potatoes are lifted in July, and are marketed locally through a central distribution co-operative called Tnuva.

2. *Dry crops* include two grains, wheat and chickpeas. There is hardly any rotation, because the crops are cultivated on marginal land and the areas cultivated vary considerably from year to year. No irrigation is required since the crops are dependent on the winter rain. The soil

is fertilised with ammonium sulphate. The crops are harvested in August, and marketed through the Central Distribution Co-operative.

3. *Orchard crops* include bananas, avocadoes and citrus fruit (including grapefruit, Valencia oranges, Washington navel oranges, Jaffa oranges, tangerines and pomelos (shaddock). Banana trees bear fruit from five to eight years after planting, and replanting can take place again only after four to five years. Large amounts of organic fertilisers plus phosphates and potash at a rate of 50 kg per dunam (494 kg per ha) are required to prepare the soil. Further applications of ammonium phosphate, potash and organic fertilisers are required during growth. Drip irrigation at 1,100 m³ per hour is required to supplement the annual rainfall. The bananas, which go for local consumption, are marketed through the central co-operative. Avocadoes require no rotation, nor do they require fertiliser prior to planting, although some organic fertilisers, nitrogen and potash may be added during growth. Sprinkler irrigation at the rate of 800 m³ per hour is used to supplement natural rainfall, and iron is added to this water. About 90 per cent of the fruit is exported through AGREXCO, the Central Export Board for fruit and vegetables, for markets in Europe. Citrus fruit is treated in much the same way as avocadoes as regards irrigation and soil dressing requirements. The sprinklers are played under the trees. No rotation is required. Some 65 per cent of the fruit is exported through AGREXCO, and another 15 per cent is canned through the MILUOT co-operative group.

Livestock. The only livestock kept are turkeys. About 350,000–400,000 poults are produced using artificial insemination and reared in the Kibbutz's own hatchery, and using a large, white broad-breasted variety of bird called "Nicholas". The laying stock is 12,000 per annum and 60,000 birds are marketed each year for meat. The average weight when marketed is 13–14 kg for a male turkey, and 6.5–7 kg for a female turkey. Another 20,000 birds are sold to neighbouring farmers at 6 weeks of age, and they fatten them to slaughter weight. The birds are shipped to the central slaughterhouse belonging to the MILUOT co-operative complex. MILUOT also supplies the feed from its modern feed-mill, developed as a result of extensive research. The birds are sold locally, or canned by yet another section of MILUOT. Some of the canned turkey and other products of MILUOT are exported to Europe.

Mechanisation. Machinery includes fourteen tractors, five sprayers, eight Michelson fruit-pickers, and the land-preparation equipment owned by the Kibbutz. Cotton-pickers, combines, and heavy earth equipment for deep ploughing, levelling, etc., are held in a co-operative pool and shared with other settlements in the area.

Income

The Kibbutz derives its income from the sources shown in Table 14.

TABLE 14

Sources of income on Kibbutz Gesher Haziv

Source		Percentage
Agriculture		70
including:		
Field crops	8	
Bananas	9	
Avocadoes	12	
Citrus fruit	4	
Turkeys	37	
Factory		16
Guest house		7
Outside employment		7
TOTAL		100

Assessment of the Kibbutz Movement

Kibbutzim are not only sound economic enterprises, but produce worthy, patriotic and concerned citizens. Many of these distinguish themselves in the upper echelons of the army, or hold important positions in government and in public institutions, a contribution out of all proportion to the size of the Kibbutz population. The intellectual and cultural achievements have also been high. The Kibbutz Movement has a fine orchestra and a renowned choir which has toured Europe quite frequently.

Perhaps the main strength of the movement lies in the fact that each member has complete financial security. Every need is provided for, albeit very simply in a less-wealthy Kibbutz. There is no need for competition, and employment is assured. In times of sorrow or rejoicing the Kibbutznik enjoys the fellowship and support of a large concerned family. As in all societies, however, the Kibbutz has to contend with laggards, braggarts and liars in its ranks. Remarkable though it may seem, the deep involvement of the members with one another once again provides the answer. With social pressure its only form of punishment, the Kibbutz maintains a happy and well-ordered society.

Chapter XV

Co-operative Agriculture

Origins and History of Agricultural Co-operation in the U.K.

THE pre-industrial world was almost entirely dependent upon subsistence agriculture, and the people were accustomed to undertaking such seasonal operations as irrigation, sowing and harvesting, as well as the herding of their livestock, with the co-operation of all the members of the community. Similarly, such people would band together for mutual aid in face of frequent disasters. The Industrial Revolution, first in Britain and subsequently in many regions throughout the world, has changed all this. Technological progress has provided greater wealth and security from natural hazards, but it has also disrupted the more orderly structure of pre-industrial society, and introduced new competitive and individualistic attitudes into society.

In the *laissez-faire* "dog eat dog" society which emerged in nineteenth-century industrial Britain, the greatest threat was to the poor and unskilled who were particularly vulnerable economically, and this gave impetus to co-operation, because, by adopting its tenets and practices, weak individuals could achieve in association with one another a level of security and prosperity denied to them on their own.

Philosophical bases for co-operation had emerged a little earlier. A Dutchman, P. C. Plockboy, domiciled in England, is credited with advancing proposals for the formation of economic associations amongst agriculturists, artisans, seamen and professionals as early as 1659. The members of such associations would pool their capital, work in the joint enterprises they established, exchange their products among themselves, share the profits, and reserve the right to withdraw their capital if and when they chose. John Bellars the Quaker put forward proposals in 1695 "for raising a College of Industry of all useful trades and husbandry"—an extension of Plockboy's ideas. He saw producer and consumer sectors working in association, capital being borrowed on the open market outside the College, and the goods produced being sold on the open market to generate profits to pay interest on the capital borrowed. Sellars influenced the better-known Robert Owen (1771–1858), whose more famous schemes included the model mill at New Lanark and the earliest trade union on a national scale, the Grand National Consolidated Trades Union, and who also guided the development of consumers' and producers' co-operatives. Most famous of the latter was the Rochdale Equitable Pioneers established in 1844, while less well known was the agricultural co-operative project at Ralahine in Ireland. Although the scheme at Ralahine collapsed when the owner's prodigality led to all his estates going into liquidation, Owen's work there demonstrated that co-operation could transform a community degraded by poverty into one hard-working, well ordered and prosperous. After the success at Rochdale a number of consumer cooperatives grew up in the north of England, and forty-four such societies attended the Northern Conference of 1851. The first specifically agricultural co-operative—a supply co-operative, "The Agricultural and Horticultural Association"—was founded in Manchester in 1867.

Despite this early flowering in Britain, however, the movement was subsequently less successful in Britain than elsewhere. Today, whereas retail and wholesale co-operative societies are well developed in Britain, it is in other Western European countries, such as Denmark, Norway and the Netherlands, and more recently in many emerging countries, that agricultural co-operative societies are most highly developed. Bonner observes that "even today almost the only substantial undertakings by co-operative enterprise in the actual running of farms in Great Britain are those of consumer co-operation, i.e. of the wholesale and retail societies", and suggests that this is because British farmers have varied so much in class and status, and that the "homogeneity of interest, indispensible to co-operation, was lacking". Early efforts to establish agricultural co-operatives amongst the peasants of Ireland certainly seem to bear out most of this.

In 1900 the co-operative movement in Britain was given new impetus by the establishment by W. L. Charleston of Newark of the British Agricultural Organisation Society. Subsequently, as the Agricultural Organisation Society (A.O.S.), it was transferred to London. By 1914 there were 274 supply societies with 30,000 members, and 129 marketing societies with 10,000 members. In the post-war slump period, however, the A.O.S. was dissolved, and it was the Co-operative Wholesale Society (C.W.S.), representing the consumer co-operative movement, which helped to sustain the agricultural societies by not only expanding its own agricultural department for the supply of agricultural requisites, but also extending credit to the societies and advising them on accountancy and business methods. In this same troublous period, in 1919, the Horace Plunkett Foundation was established to "promote systematic study of the principles and methods of agricultural and industrial co-operation". Now, under the name of the "Plunkett Foundation for Co-operative Studies", it continues this task, and indeed this study owes much to the help of the Foundation and its former chief executive officer, F. H. Webster.

In 1930 the Plunkett Foundation took the initiative of calling a conference of co-operative societies in London to find out ways towards wider co-operation between them by setting up a Central National Body. The delegates, members of the National Farmers' Union (N.F.U.) and some members of the Co-operative Union (C.U.), were fearful lest, in setting up such an organisation, they might damage these affiliations. Instead of setting up a central body, they wanted greater representation for the individual societies with the N.F.U. and the C.U. The C.U. offered to set up an agricultural section, and the N.F.U. offered a number of limited concessions. The proposal for a central body was shelved, the conference having achieved for the agricultural co-operative societies the means for closer collaboration with the N.F.U. and C.U. in the future. Since the Second World War, this has culminated in the establishment of the Agricultural Co-operation and Marketing Services (A.C.M.S.) in 1972, based upon the amalgamation of the N.F.U.'s Market Development Unit with the Agricultural Central Co-operative Association, itself a product of an earlier merger between the Agricultural Co-operative Association, founded in 1946, and the then Farmers' Central Organisation of the N.F.U. The A.C.M.S. and other central societies in Wales, Scotland and Ulster make up the Federation of Agricultural Co-operatives, F.A.C. (U.K.), which also represents U.K. agricultural interests with C.O.G.E.C.A., the E.E.C. organisation of agricultural co-operative societies which meets in Brussels.

During the period between the wars membership of co-operatives dwindled until in 1938 it had fallen to about half that of 1918. The post-war recovery has been such that in 1966–7 there were 535 agricultural co-operative societies in the U.K. with 418,980 farmer members, including 273 English societies with 263,841 members. More up-to-date figures are included in Table 17 on p. 261.

In 1967 parliamentary parties of all persuasions united to support the establishment of the Central Council for Agricultural and Horticultural Co-operation. The Council's terms of reference are " to organise, promote, encourage and co-ordinate co-operation in agriculture and horticulture", and its work involves providing legal, financial and marketing advice to farmers, undertaking training and promotional activities, and helping in the preparation of E.E.C. grant applications. The last of these has come to the fore since 1973, when the U.K. joined the E.E.C., as a result of which farmers have undertaken amalgamations and improvements, and co-operative societies have been encouraged to develop their marketing activities, particularly those for potatoes, vegetables, livestock and meat.

The main co-operative functions are those concerned with agricultural supply, "services" such as grain-drying and storage, with or without milling and mixing, as well as irrigation, pest-control, etc., production, and marketing. In practice most British societies may be classified as falling into the following groups: (*a*) multi-purpose societies, which have usually grown out of supply co-operatives, but which have acquired service and market functions; (*b*) marketing societies; and (*c*) service and miscellaneous societies. All such societies have characteristically incorporated the following common principles into their constitutions.

1. The trading surplus, or profits accrued to the company by trading, must be redistributed to the members in shares directly proportionate to the amount of business they have done over the same trading period with the society.
2. Interest payable on loans or shares in a co-operative society is strictly limited. Co-operative society shares themselves do not appreciate in value.
3. There is no restriction upon membership, other than that members are expected to be actively engaged in farming, to produce commodities of an acceptable standard, and to display a reasonable commitment to the society in terms of regular trading, and a capital investment roughly proportionate to that trade.
4. Whether their commitment, in trade or investment, is large or small, however, all members have an equal share in the policy formation and management oversight of their society, through the principle of "one man—one vote".
5. Agricultural co-operative societies may vary in their legal status, being constituted either as societies under the Industrial Provident Act 1965, or as limited companies under the Companies Acts 1948 and 1967. Producer co-operatives only may also opt to function as partnerships, in which case only a written constitution is required, and the formal registration is dispensed with.

Agricultural Co-operation in the U.K. Today

The present state of agricultural co-operation in the U.K. is summarised in Tables 15 and 16. These are based upon figures for 1973–6 obtained from F. H. Webster, *Yearbook of Agricultural Co-operation*, 1976, pp. 63 and 67–8. and J. A. E. Morley, *British Agricultural Co-operatives*.

To illustrate the practical achievements of agricultural co-operation in Britain the following two studies of horticultural co-operative societies in England have been included.

Nursery Trades (Lea Valley) Ltd.

The early 1920s were a period of economic recession which afforded severe difficulties to horticulturists. It was against this background that the first society to be considered was set up. Nursery Trades Ltd. was established in 1921 to supply local growers with requirements such as

TABLE 15

Turnover and membership of co-operatives in the U.K.

	Co-operative societies (from the Friendly Society Annual Returns, 1974–5)	Co-operative companies (figures for 1973)
Total turnover	£683.57 m.	£139 m
Proportion of total turnover in		
England	71.5%	99%
Wales	5.7%	1%
Scotland	17.7%	—
Northern Ireland	5.1%	—
	100.0%	100.0%
Proportion of total turnover in		
Supplies	60%	17%
Marketing	40%	55%
Services	—	28%
	100%	100%
Number of organisations	415	186
Number of members	313,400	29,000

TABLE 16

Estimated share of national totals of supplies and sales handled by co-operatives in the U.K.

	Percentage
Supplies of:	
Fodder	15
Fertilisers	18
Sales of:	
Cereals	12
Eggs	17
Fat-pigs	4
Fat cattle and calves	9
Fat sheep and lambs	8
Wool	31
Potatoes	9
Top fruit	20
Horticultural produce	9

coal, glass, formaldehyde and fertilisers, and is indeed one of the oldest horticultural supply co-operatives in existence. In 1937 a marketing society—Glasshouse Growers Sales Ltd.—was established for tomato and cucumber production at Waltham Abbey. In 1947 the two societies

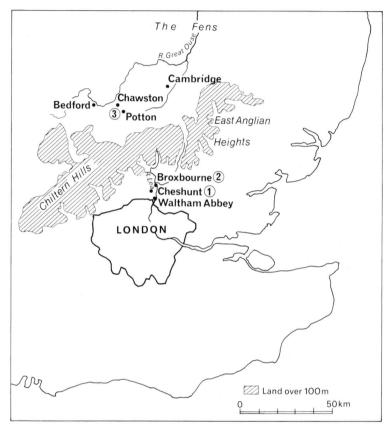

FIG. 106.—*Location of co-operative societies, estates and growers described in text.* 1. Nursery Trades (Lea Valley) Ltd. at Cheshunt. 2. Thomas Rochford and Sons Ltd. at Broxbourne. 3. Land Settlement Association Estates at Chawston and Potton, Bedfordshire (*see also* Chapter XIII).

were amalgamated to form Nursery Trades (Lea Valley) Ltd., with headquarters at Cheshunt (*see* Fig. 106). From then on it ran a requisites department mainly to provide for the needs of growers in the Lea Valley, where in 1952 there were 445 ha under glass. Over the years this has declined to 182 ha, and so to maintain its viability the society has had to provide requisites over a much wider area. There is a much larger membership now of 1,162 in an area extending as far north as

Suffolk, as far south as Surrey, and as far west as Oxford. The marketing department, which specialised latterly in cucumbers, was discontinued in 1974–5, leaving the society a supply co-operative, as originally conceived.

In 1954 the society had a membership of 452, and its total share capital and reserves stood at a little over £36,000. The society today has 1,162 members and combined share capital and reserves of nearly £346,000. While much of this increased capital holding represents the effects of inflation on money values, there is still evidence of substantial growth in the volume of business and total turnover.

A member is required to purchase a minimum of twenty-five £1 shares to join the society and is expected in addition to build up a shareholding proportionate to his holding of glass. Provided this condition has been met he is entitled to receive share interest, and a deferred rebate for prompt settlement of his accounts. If there is a profit balance after these items have been met, and it is deemed prudent, a further trading rebate on members' purchases of eligible goods may also be paid out.

At the A.G.M. the members elect eleven of their number on to the Board of Directors, presided over by the Managing Director. The Board meets four times a year to decide on matters of policy and watch over society affairs. The day-to-day operation of the society is, however, handled by the management and salaried staff.

The main services provided by the society include supply and delivery of a wide range of requisites to growers and top fruit producers in the Lea Valley, Home Counties and East Anglia. These requisites include fertilisers, containers, etc., which have been bought in bulk and are supplied to growers at the lowest possible prices.

In addition, two associated companies exist to provide specialised services to members. Lea Vale Fuels Ltd. buys fuel oil and gas oil for sale to members. In the year ending June 1977, 32 million litres of oil worth £1.667 m. were sold. Lea Valley Growers' Transport Ltd. secures bulk carriage rates on British Rail Passenger Services for members.

Thomas Rockford & Sons Ltd., of Turnford Hall Nurseries, Broxbourne, Hertfordshire, is one of the members of the Nursery Trades (Lea Valley) Co-operative. It is an old and established supplier of house plants and dates back to 1877, when its premises were on the site now occupied by the Hotspurs Football Club, but in 1881 it moved to its Broxbourne site (*see* Fig. 106). The significance to our present study is that Rochford and Sons, because of the enormous scale of its horticultural operations, is a major customer of Nursery Trades. There are about 15 ha of glasshouses, with 12 ha on the main site and 3 ha on two other sites within a radius of 21 km. This vast area of glasshouse requires provisions of fertiliser, soil nutrients, and plant containers of various kinds and sizes. In addition there are two boiler houses, No. 1

with six oil-fired boilers erected in 1966, and No. 2 with two further boilers, which between them consume 11.4 million litres of heavy fuel oil per annum.

The Land Settlement Association

In the 1930s there were many unemployed in the mining and heavy industrial regions of the country, and a number of voluntary bodies such as the Society of Friends got together and proposed a scheme by which land purchased jointly by private subscription and government contribution could be used to provide such men with smallholdings. The following four principles were to govern such help.

1. Assistance was to be given only to group settlements.
2. Co-operative methods were to be adopted for the supply of requisites and the marketing of produce, and the general organisation of settlements within the scheme.
3. The men and their wives were to be selected, with preference given to those who had successfully cultivated allotments.
4. Training and supervision should be given to candidates, but no scheme was to be considered unless it was large enough to justify the appointment of a full-time supervisor.

The scheme gained government approval, and the Land Settlement Association was registered under the Industrial and Provident Societies Acts in the latter part of 1934. The first estate, described subsequently, was actually given to the new association. In due course estates sprang up all over the country, with increased government support. The aim was towards a "three-legged stool" economy, based on pigs, chickens and horticulture, to provide maximum economic stability.

There were three major developments in policy in the changed circumstances of post-war Britain.

1. The estates were taken over by the government, and the L.S.A. was reconstituted as an agent to the Ministry of Agriculture, to provide the centralised services and to administer the estates.
2. The "three-legged stool" type of economy was abandoned in favour of specialised horticulture, which was now far more profitable.
3. The policy of accepting unemployed industrial workers was replaced by one requiring candidates to be experienced and/or qualified in agriculture, and to have a minimum of capital or credit available to pay for initial investment in their holding.

The Land Settlement Association today. The L.S.A., with its head office in London, is directly responsible to the Ministry of Agriculture. It is not, strictly speaking, a co-operative organisation at all, even though it is registered under the Industrial and Provident Societies Acts, because it is made up mainly of independent people interested in the work of the association. However, while the relationship between the tenants and the L.S.A. is closest to that between tenant and landlord, and while the contractual agreement under which the tenant is bound to use the central services provided on the estate is seen by purists as the very negation of agreements entered into voluntarily by co-operators, there is none the less a democratic system of access between tenants and the L.S.A., and a community of interest between tenants on an estate, which certainly reflect some of the spirit of co-operation.

On a purely local level the tenant consults his estate manager about any difficulties, and if the problem is easily settled the matter ends there. Failing this, the tenant takes the matter to the Tenants' Estates Committee, who may carry it back to the estate's manager. If, however, the problem is of more than local concern, it may be dealt with higher up the system. The Chairman of the Tenants' Estates Committee is also *ex officio* a member of the Central Tenants' Council, and this body in turn nominates certain of its members to the Minister of Agriculture for appointment to the Executive Committee. Furthermore, the Executive Committee and the Central Tenants' Council make up the Central Joint Committee and both these bodies operate in close liaison with the Minister himself. Thus the tenant who elects one of his colleagues to the Tenants' Estate Committee, through him has a voice in the highest assembly, since the Executive Committee appointed by the Minister includes the Chairman and Vice-Chairman of the L.S.A., with four independent members, and four tenant members selected for appointment by the Minister from among the elected Chairmen of the Tenants' Estates Committees.

The L.S.A. estates, of which there are ten in all, with about 540 growers or tenants, vary as regards the size of holdings, service facilities offered, over-all size of the estate, and, of course, their exact form of land-use, since this depends not only upon local geographical conditions, but also upon local market conditions and tenants' own preferences. In general, an average estate has some fifty holdings of about 2 ha each. There is a three or four bedroom house on each holding with mains and services, and store-sheds and glasshouses can be taken over by an incoming tenant for an agreed valuation. The tenant, who has to show evidence of five years' full-time horticultural or agricultural experience, but which can include up to three years at an agricultural college or university, and ability to provide 25 per cent of the required capital (the other 75 per cent can be borrowed), also has to agree to use the centralised services of the estate. Production on all the estates is almost entirely by intensive cultivation of horticultural crops—mainly under glass—although there are some estates where pigs and chickens are still kept. The centralised services include propagation of plants, bulk purchase of commodities required, provision of machinery and specialised equipment, and the packing, grading and marketing of produce under the L.S.A. trade mark. The estate manager is responsible to the general manager in the head office in London for the efficient operation of these services.

The Potton Estate. As an example of such an estate, a brief account is given here of the oldest of the L.S.A. estates, at Potton. For administrative and marketing purposes it is amalgamated with another estate at Chawston (*see* Fig. 106) and both are administered from Wyboston, a small Bedfordshire village 14.5 km from Bedford and 29 km from Cambridge on the A1 trunk road. The total area of the two estates is 298.25 ha and these include 100 smallholdings of which 93 are currently occupied, each with a three or four bedroom house, a store building and an average 2 ha of land. Chawston actually averages 2.2 ha, while Potton averages 2.5 ha. Most of the estates' land is given over to horticultural produce, with 20 ha under glass, of which 60 per cent (12 ha) is heated. In addition polythene tunnels are being increasingly used. Glass occupies 0.2–0.6 ha of each holding. Products under glass include lettuce, tomatoes and self-blanching celery, and lettuces and self-blanching celery are chief amongst the open-air crops. Output of the combined estates in 1977 was worth £1.141 m., the chief items being:

256,242 cases of tomatoes (each weighing 5.4 kg)
445,129 boxes of lettuces
 88,610 cartons of self-blanching celery
  6,642 boxes of peppers

Centralised services include a propagating plant with 0.87 ha of glass employed in raising glasshouse plants for growers, who also include some private nurserymen and ex-L.S.A. tenants. This department raises 620,000 tomato plants, 1.5 million self-blanching celery plants, 6 million lettuce plants, 125,000 tomato seedlings, 1.2 million celery seedlings and 30,000 pepper plants per annum. There is a packing station which grades and packs the produce of both estates and distributes it to wholesale markets and self-service stores under the L.S.A. trade mark. The weekly returns for each grade of product are averaged and each grower receives the averaged price proportionate to the volume of produce of each grade he has contributed.

Service cultivations are carried out, out of doors or under glass, using the estate machinery and tractor. Estate transport lorries distribute produce to markets or collect containers, fertilisers and other growers' requisites from suppliers. The stores stock a range of such requisites— seeds, equipment, glasshouses, irrigation plants, heaters, insecticides, fungicides—and also arrange contract services such as soil fumigation and biological control on behalf of growers. Some ex-L.S.A. and other growers also use the stores. The estate accountant at the office administers the estates, keeping the growers' monthly accounts, yearly profit and loss accounts, and contracts for services, etc. The cropping programme—crops to be grown each season—is reviewed annually in the light of L.S.A. marketing policy and in consultation with growers. As regards the over-all management, policy matters and marketing plans are determined at head office. The L.S.A. is responsible to the Minister through his Executive Committee. The resident estate managers meet regularly with the growers' representatives at a monthly Estate Committee meeting.

A map of the Potton Estate is shown in Fig. 99 in Chapter XIII, where J. W. Parker's horticultural holding is described (*see* pp. 225–228). Mr. Parker has given his own impression of changes on the Potton Estate over the past decade. After observing that growers once worked with "pigs, poultry and outdoor market gardening", he writes that "over the past 12–15 years the swing has been to glasshouse salad crops. I have been here for ten years and started with some glass and outdoor cropping. As I have built up the glass, so the outdoor crops have diminished to negligible proportions."

The Co-operative Movement in Western Europe

Nine of the countries of Western Europe are members of the European Economic Community, or Common Market, which will be the subject of fuller consideration in Chapter XVII. This chapter has already considered one of these countries, the U.K., from the point of view of agricultural co-operative development, and another member, Denmark, is examined on pp. 263–9. In both these countries the E.E.C. is concerned to promote greater efficiency of agricultural marketing. Since it was established in 1957, the E.E.C. has expanded and moved towards economic integration, and although such integration has been uneven, nowhere has it been more fully implemented than in the agricultural sector, a process in which the agricultural co-operative movement has played a part.

In 1975 E.E.C. agriculture contributed £41.62 m. or rather less than 7 per cent of the gross domestic product of the Community. Yet in the same year it still accounted for 8.7 per cent of the total Community employment, although this over-all average conceals disparities such as 3 per cent in the U.K., and 20 per cent in Ireland. Despite the fact that co-operatives have been proved advantageous to regions with small farms, which are characteristic of several E.E.C. countries such as Italy, there is still scope for extending co-operation. This is also true in the U.K., where the farms are well above the average size for the Community.

Not only does the E.E.C. possess considerable economic potential, but it already ranks high in the scale of world economic powers, as shown in Table 17. Despite its high agricultural production relative to so many other regions of the world, the high population of the E.E.C. also makes it an over-all deficit producer. Although it enjoys near self-sufficiency in some products, and a surplus in just a few, it is obliged to import in order to meet deficits in a number of others.

As regards agricultural production inside the Community, the Common Agricultural Policy (C.A.P.) aims to ensure increased production, adequate living standards for producers and staple supplies and fair prices for consumers. Elaborate legislation has been formulated to promote these ends. In addition to E.E.C. regulations concerned with consumer protection, there is also a canon of national legislation, framed

TABLE 17

Ranking of E.E.C. and world economic powers

	Population (millions)	G.D.P. (million E.U.R.s*)	Cereal production average 1972–4 (million tonnes)	Meat production 1974 (million tonnes)	Milk production 1974 (million tonnes)
E.E.C.	256.6	835,600	105.8	20.4	101.9
U.S.A.	210.4	1,038,000	219.0	24.8	52.4
U.S.S.R.	247.5	N/A	185.5	14.7	92.3
Japan	108.4	331,700	60.6	2.2	4.8

NOTE: After Gregory C. Tierney, "Farm co-operation in the E.E.C.", *Yearbook of Agricultural Co-operation*, 1977, p. 52.* "E.U.R." (sometimes rendered E.U.A.—European Unit of Account) is the monetary unit used in pricing the Community budget. Each unit is worth 0.888671 grams (35 units = 1 oz.) of fine gold—values equivalent to the gold content of the U.S. dollar in 1944.

in response to consumer pressure groups, and designed to regulate quality, packing and labelling, and to establish legal liability for defective products. Although such legislation is not specifically directed towards co-operative societies, it affects them as marketing and processing organisations along with private companies. Problems involved in conforming to C.A.P. and national regulations have posed a challenge to the over-all membership of co-operative societies, as well as prompting an increase in the scale of enterprise which has been achieved by both amalgamation and increased inter-cooperative trading.

Co-operatives have assumed a predominant position in the marketing of milk, and control a significant proportion of most produce markets, particularly those concerned with cereals, fruit and vegetables, as indicated in Table 18. However, there is very little evidence that this has been brought about by membership of the E.E.C. which takes a fairly neutral stand in the matter of types of business organisation adopted within the agricultural industry. Thus the E.E.C. has done little to promote cross-frontier amalgamation or collaboration between societies, although it has exempted co-operatives from restrictions designed specifically to check such monopolistic practices as "price-rings", showing recognition that co-operatives serve the best interests of the Community. The E.E.C. is also committed to setting up "producer groups" for fishermen, horticulturists, and sericulturists in particular, although apparently showing indifference as to whether such groups should be co-operatives, limited companies or partnerships. An inherent danger of this development is that by evolving their own marketing systems, such groups could undermine the established marketing co-operatives, and this matter is of some concern in C.O.G.E.C.A.

The European Agricultural Guidance and Guarantee Fund (known by its French initials, F.E.O.G.A.); under its "guidance" function, has

TABLE 18

Proportions of produce markets controlled by co-operatives in E.E.C. countries (percentage)

	Milk	Beef	Pig-meat	Fruit	Vegetables	Cereals
Germany	78	19	20	26	36	52
France	46	15	50	40	30	70
Italy	35	5	5	46	5	15
Netherlands	90	20	30	82	85	60
Belgium	65	small	15	35	50	15
Luxembourg	90	22		30–35	—	70–85
U.K.	—	9	5	15	9	20
Ireland	88	28	32	16	18	30
Denmark	87	60	90	55–60	50	40

NOTE: After Tierney, *op. cit.*, p. 56 (*see* NOTE on p. 261 above) and based upon statistical estimates of the European Commission.

approved development grants to both producer and co-operative marketing schemes. Both types of scheme are able to offer assured security of supply in support of their economic viability. Whereas, however, since 1964 such schemes have been approved individually, under new regulations grant aid will require that the project supported should fit in well with an over-all national strategy framed by the national government concerned. In general the co-operative movement is in favour of such schemes, confidently believing in its own ability to fit into any scheme which helps to ensure more effective use of the Community's financial resources.

Reference has been made to tendencies towards increased inter-co-operative trading, even at an international level. C.O.G.E.C.A. has prepared a directory of co-operative societies interested in such trade amongst the member states. An even more exciting tendency is towards the development of direct trading links, promoted by Euroco-op—Europe's consumer co-operative organisation—between consumer and producer co-operatives. Such direct links could, by eliminating the middleman, bring substantial benefits to producer and consumer alike, and the Community might well provide the essential framework for such a development.

Danish Agriculture and Co-operation

Although there has been a gradual tendency since the eighteenth century for farms in Denmark to grow, the last few years have accelerated this process. Whereas in 1880 there were 180,000 farms and smallholdings of less than 0.5 ha, in 1976 this number had fallen to 124,000, and is expected to drop to 80,000 by 1985. Of the total land area in Denmark, 68 per cent or approximately 2.89 million ha is used for agriculture, and of this, 90 per cent is arable, and a further 9 per cent under permanent grass. As a result of the amalgamation of farms since 1960, when certain outmoded laws restricting the process were rescinded, the average size of farm-holding has risen from 15.8 ha to 23.4 ha in 1976, although this average conceals the surprising 45 per cent of farms of less than 15 ha remaining. There are regional variations too, with a smaller average farm size in Funen (20 ha) than in South Jutland (31 ha).

Although many owner-farmers rent additional land, only 3,100 tenant farmers occupy 3.3 per cent of the total land farmed, while the remaining 120,900 farms are privately owned, and occupy 96.7 per cent of the land.

The Danish Co-operative movement

While a farmer's output is related to his skill and the productive capacity of his holding, his income depends more upon the margin between the costs of his inputs and the price he gets for his outputs. His income, therefore, and as a result the ultimate prosperity of Denmark's economy, as well as the condition of international market prices, depend upon factors outside the individual farmer's control. In order to share in the control of these forces external to his actual farm operations, the Danish farmer has for a long time joined with his fellows in the establishment of supply and marketing co-operative societies through which he can help to influence matters.

The Danish co-operative movement first came into existence as a result of events which had also affected Denmark's chief trading partner, Great Britain. In the early nineteenth century Denmark was exporting wheat to supply Britain's expanding industrial market. In the 1860s, however, the opening up of the Canadian Prairies by rail and water communications gave Canada access to the same market (*see* Chapter IX) and Denmark could no longer hope to compete. The

Danes sought to develop animal products as their principal export. The small size of Danish holdings, the perishability of dairy produce and the need for organised processing and marketing all encouraged the Danes to search for some form of co-ordinating and controlling organisation. This they found in the already existing Thisted Co-operative Consumers' Society which had been established in North Jutland in 1866, itself based upon the earlier Rochdale Equitable Pioneers Society established in Britain as early as 1844. Whereas in Britain such consumer co-operatives had grown up mainly in the towns, those in Denmark caught on most rapidly in rural areas, and provided inspiration for the setting up of the first specifically agricultural co-operative in Denmark at Hjedding, West Jutland, where a co-operative dairy was established as early as 1882. It rapidly proved successful and was widely copied. The first co-operative bacon factory was founded in 1887, and by the 1890s societies were seeking to co-ordinate their purchasing operations, which prompted the establishment of the Danish Co-operative Wholesale Society in 1896. Co-operative egg exports began in 1895, and cattle exports by the turn of the century. At this time fodder and fertiliser purchase associations emerged.

Soon national societies, to which local societies were affiliated, grew up to co-ordinate such activities as fodder and fertiliser purchase, and dairy and bacon production. Then in 1899 the national organisations established the Central Co-operative Committee to co-ordinate their common aims, publicise their ideals, and plan a co-operative bank. This committee grew to become the Federation of Danish Co-operative Societies. Clearly the entire Danish co-operative movement grew in response to the needs expressed by the farmers themselves.

Naturally, in an economy which is very much based on agriculture, the central government must also be concerned with the well-being of the industry. So it was that in 1919 the Federation of Danish Co-operative Societies, along with the Federation of Danish Farmers' Unions and the Royal Danish Agricultural Society—a learned philanthropic society which had been established in 1769—set up the Agricultural Council to provide collective representation for the agricultural interests generally in their dealings with central government concerning matters of domestic and international concern. Through evolution, the Agricultural Council has become far more representative of the industry, with delegates from the Danish Smallholders' Union and the Export Boards and Committees established by the government in 1932, but since 1950 regulated by the agricultural interests involved, in addition to the agricultural interests already mentioned, meeting to negotiate with the E.E.C. through C.O.G.E.C.A.

It is appropriate here to explain the nature of the Export Boards and Committees before referring to the relationship between local co-operative societies and the farmers themselves. During the economic depression of the 1930s agriculture was itself depressed. The Danish Ministry of Agriculture pushed through crisis legislation in 1932 for the establishment of Export Committees in order to provide for "closer working between producers, processors, and exporters" within each of several product groups. The number of such groups was subsequently increased. Later, in the era of expansion following the Second World War, the work of the committees came under the control of the agricultural interests involved. Today ten Export Boards and Committees continue to co-ordinate the activities of producer, processor and market organisations within their particular product group—working to promote the quality of goods, the most favourable prices and improved sales. Others exert direct influence on supplies and price policies, or undertake sales themselves. Under E.E.C. provisions they are empowered to acquire financial support through production levies. These committees, which are *ex officio* members of the Agricultural Council, include the Dairy Industry's Butter Export Board and its Cheese Export Board, the Danish Bacon Factories' Export Association (ESS-FOOD), the Danish Livestock and Meat Board, the Poultry Export Committee, the Egg Export Committee, the Potato Export Board, the Seed Section Export Committee, the Danish Fur-breeders' Association, and the Federation of Danish Sugar-beet Growers' Association.

Place of co-operative societies in the agricultural economy

If we consider the entire productive chain from grower to consumer, the co-operatives do not dominate it throughout, but their presence may be said to provide necessary competition which serves to ensure the efficiency of the private sector, so that producer and consumer alike receive adequate services at reasonable cost.

Co-operative enterprises provide 50 per cent of fodder supplies, 45 per cent of seeds, and 43 per cent of fertilisers. In the early stages of processing farm produce, the co-operatives generally control the major share, with 91 per cent of bacon pigs, 87 per cent of milk, but only 60 per cent of cattle; while in the subsequent processing and exporting stages co-operative processors handle 25 per cent of exports, and co-operative marketing associations a further 50 per cent of exports. Amongst the largest of these agricultural marketing organisations are ESS-FOOD (*see* above) and BUTTERDANE, which controls 90 per cent of all butter exports and counts the majority of dairies amongst its members. (*See also* p. 268.)

Changes in Danish farming

Increased specialisation in production means that the mixed farmer is no longer typical of Danish agriculture as a whole, and in these circumstances it is better to describe farming in a more generalised way.

To illustrate trends towards specialised production, the following quotation from *Agricultural Co-operation in Denmark* by F. H. Webster (Plunkett Foundation for Co-operative Studies, Occasional Paper No. 39, 1973, pp. 144–5) is apposite.

Farmer Jenson lives in Vraa, North Jutland, where he farms seventy acres [28.3 ha], which has been in his family for seven generations; his wife is also from a farming family. In 1954 Farmer Jenson was in his middle forties, and ran the farm with a cowman and his daughter, with occasional help from a neighbour, who had a smallholding, and was glad of the opportunity to augment his income. The farm had twenty milking cows with heifers and calves, 40–80 pigs, and several hundred poultry. The livestock was fed from the farm as far as possible, and seed corn was the only crop sold. Fifteen acres [6 ha] were permanent grass, ploughed up every six to ten years; twenty-eight acres [11.3 ha] of cereals (mostly barley); thirteen acres [5.3 ha] of sugar-beet and swedes, and fourteen acres [5.7 ha] of rotation grasses. Farmer Jenson was a member of the following co-operatives: dairy, bacon factory, poultry packing, cattle marketing, fur auction (he ran a fur farm with two other farmers), feeding stuffs, fertilisers, village shop, Jutland 'Farmers' Credit Society and the village savings bank. He also insured himself and the farm through co-operative insurance.

In 1973 Farmer Jenson is as active as ever, but because of the scarcity and cost of farm labour, he and his wife run the farm alone. Farming has become more specialised, and the farm now concentrates on poultry, pigs and barley, of which about sixty acres [24 ha] produces about 120 tons per year. This is marketed through D.L.G. [Danish Co-operative Farm Supplies] with which the Jutland Co-operative has become amalgamated. The farm has no dairy or beef cattle; Mrs. Jenson is particularly interested in developing poultry production, however, and about 17,000 chickens are marketed five times each year to the Danish Poultry Association, which is now centralised on a nationwide basis. Farmer Jenson is still a member of his Farmers' Association, Hjørring Amts Land-Economics Society, which provides advisory services and ... organises the pig auctions in Hjørring for those animals which are not marketed to the bacon factory.

Hjørring has a dairy and bacon factory, which both service much larger areas as a result of the rationalisation schemes ... Farmer Jenson's attitude to the rationalisation schemes is pragmatic: "Centralisation works well on the whole, it benefits the farmer, but something of the old spirit is lost." ... the co-operatives now emphasise the importance of management rather than membership involvement ...

Fuller details of Farmer Jenson will be found in *Co-operation and the Danish Farmer* by J. A. B. Hamilton (Occasional Paper No. 11 (1956) of the Horace Plunkett Foundation, now called the Plunkett Foundation for Co-operative Studies). It will be seen how a diverse operation, which in 1954 demonstrated almost a cross-section of the Danish farming economy, has in less than twenty years been reduced to one specialised in "poultry, pigs and barley" with plans to intensify poultry production in the future.

The following generalised description also emphasises the growing diversities.

Denmark is composed just over half by the Jutland Peninsula and the remainder by 500 islands, including 100 that are inhabited. No place is more than 52 km from the sea, while the average altitude throughout the country is 30 m, and so in general terms the climate may be expected to be equable and homogeneous, even though western areas are rather wetter and milder in climate than eastern ones. Mean temperatures are about 0°C in January, and 16°C in July, while the annual rainfall total is about 665 mm. The soils are generally of glacial origin, composed of varying proportions of clay, sand and gravel. The glacial outwash soils of western regions contain sand and gravel predominantly, while the ground moraines of the eastern areas left soils with a large proportion of clay.

Although in the past farms were very small, the present average is 23.4 ha and is likely to increase still further in the foreseeable future. As we have already noted, most farms at one time were "mixed enterprises" with pigs and cattle, as well as arable crops grown for fodder. Today, although about half of the average sized farms still keep pigs and cattle, this number is declining and an increasing number are taking to specialisation—usually in pigs on the smaller holdings, and in cattle on the larger farms. Even amongst cattle farmers increased specialisation is developing between beef and dairy production, and amongst pig farmers between keeping weaners or fat-pigs, while poultry farmers vary between those who specialise in broilers and those who produce eggs. More than half of the very small farms, and 17 per cent of all farms, keep no livestock at all, and this tendency is also on the increase. Even on larger farms, i.e. those of 30 ha or more, which usually still keep cattle, pigs or both, there is a growing area of land now given over to crops, particularly cereals, mainly for use as fodder, so that small surpluses of cereal actually occur. An increased proportion of the larger farms now only produce arable crops, in accordance with a general trend towards fewer, but larger, herds. From the above it will be seen how it is increasingly difficult to generalise about Danish farms.

Another recent change has been the decline in the popularity of the Red Danish breed of cattle in favour of the Danish Black and White breed which has a somewhat higher milk yield, while the smaller population of Jerseys, kept for their richer milk, has remained a constant 18 per cent of the total cattle population. The Danish Landrace or bacon pig, a product of eighty years of improvement, still retains its absolute sway.

Danish farms are highly mechanised, particularly on the larger holdings, since most of them are operated solely on family labour—in 1976 only 13 per cent of all farms employed any full-time farm workers. The average age of farmers is rising. Whereas in 1950 it stood at 48 for all sizes of farm, it now stands at 52 years, and this inevitably contributes to the trend towards increased mechanisation, and replacement of more labour-intensive livestock with arable production.

Co-operative and other organisations with which farmers deal

Which societies a farmer chooses to join depends upon his production specialisms. He can choose to belong to any number of such societies whose services will prove useful to him. First, there are the supply co-operatives which provide, between them, 50 per cent of feedstuffs, 43 per cent of fertilisers and 13 per cent of the machinery purchased by Danish farmers. These include D.L.A.M. (Danish Co-operative Machinery Supplies), D.L.G. (Danish Co-operative Farm Supplies) and F.A.F. (Funen Co-operative Feeding Stuffs). The Danish Seed Trade Association (D.L.F.) specialises in the production and marketing of agricultural seeds. The cattle sale societies handle the marketing of live cattle. There are few comparable societies for pigs because the co-operative bacon factories collect the farmer's fat-pigs straight from the farm—and also cattle from farmers or from markets for slaughter. Exports of bacon go mainly to the U.K., and are handled by ESS-FOOD. Co-operative dairies collect milk from the farms for making

TABLE 19

Production, consumption and export of livestock products in Denmark, 1976

Commodity	Production ('000 tonnes)	Home consumption (percentage)	Exports (percentage)
Butter	141	66	27
Cheese	154	75	28
Eggs	72	11	80
Beef and veal	268	54	36
Pig-meat	771	69	29
Poultry (dead-weight)	117	55	43

butter, cheese, etc., and sometimes sell their by-products back to their members for pig-fattening. Eggs are marketed, mainly on the home market, by the Farmers' Co-operative Egg Association, while broiler production is handled by DANPO A/S. These various co-operative enterprises handle 45 per cent of the seed, 60 per cent of the cattle, 91 per cent of the bacon pigs, 87 per cent of the milk, 90 per cent of the butter, 59 per cent of eggs and 55 per cent of broiler chicken production. Another society is G.A.S.A., the Market Gardeners' Co-operative Marketing Association. There are also co-operative canneries, sugar-beet factories, and plants which make potato meal, not to mention the retail co-operatives which are patronised by much of the rural community.

Other organisations to which a farmer may belong or with whom he may deal include milk recording societies, which not only keep milk yield records but also provide guidance on the diet regulation and selective breeding of cattle; pig recording societies, which provide a similar range of services to pig owners; artificial insemination societies, which

operate bull and boar stations; breeding centres, which undertake pro-
geny-testing; and finally the Farmers' and Smallholders' Unions, which
provide educational facilities and advisory services, help keep yield
records and farm accounts, and give guidance to farmers with management problems.

Statistics of production and exports in Danish agriculture

With only 7 per cent of her population engaged in agriculture, Denmark
still produces three times her domestic requirements of food, which
leaves 65–70 per cent of her output for export. These same exports
constitute 30 per cent of the national total.

Of Denmark's arable production, 85 per cent supplies her enormous
domestic needs for fodder, and provides for self-sufficiency in agricultural seed, domestic and industrial grains, sugar and potatoes, still leaving a balance of 15 per cent for export.

The bulk of Denmark's agricultural exports are from her livestock,
and Table 19 gives statistics for production, consumption and exports
of these products. It will be noted that the two columns for home consumption and exports do not total 100 per cent. The discrepancies may
be explained by balances held in stock, released from stock, or retained
for an alternative use, e.g. eggs for hatching.

Trading partners fall into three main categories: E.E.C. countries,
EFTA, and the rest of the world. The E.E.C. takes 75 per cent of Danish
exports; EFTA, including Norway, Sweden, Switzerland, Finland, Por-
tugal and Austria, between them take another 6.4 per cent; leaving
nearly 19 per cent of exports going to several countries, the most important of which are the U.S.A. (6.7 per cent), Japan (3.4 per cent) and
the U.S.S.R. (1.1 per cent).

Chapter XVI

Factory Farming

Development and Characteristics of the System

EVER since prehistory the population of the world has been steadily growing, as man has learned how better to conquer his environment and so been able to provide for an ever-increasing number of people. This growth, however, has been at an accelerating rate, particularly in the past 250 years, a period which has also seen the growth of urban industry with all its implications for agriculture. At one time most of the population lived off the land, eating the foodstuffs they laboured to produce. Industrialisation brought concomitant urban expansion, creating a new landless urban population wholly dependent upon agriculturists elsewhere for their food since they could no longer grow their own. Thus industrial growth served to provide agriculture with a profitable market for large surpluses, and introduced farmers to new agricultural technologies which made production of such surpluses possible. A new problem, however, has arisen, partly because accelerating population growth has engulfed a great deal of agricultural land, but also because, as living standards have risen, people have come to regard food not as the first of their priorities, but as a basic right that can be taken for granted. The rural labourer of about 1860 spent approximately 12s. (60p) of his weekly wages of 14s. (70p), i.e. 85 per cent, on food, whereas his counterpart in the mid-1970s was unlikely to spend more than £25 of his weekly wage of £60, i.e. about 40 per cent (Sir Frank Markham, *History of Milton Keynes and District*, Vol. 2, White Crescent Press, 1975). Today people have greater expectations as regards the quality of life, in terms of holidays, cars, television and modern household amenities, and expect to spend relatively less on food. The farmer is therefore under pressure to produce more and comparatively cheaper food on the declining area of land remaining to him.

Such trends have already reached disturbing proportions in just a few very built-up regions, such as parts of the north-east quadrant of the U.S.A. and some of the more densely populated regions of north-west Europe, including parts of south-east England.

The factory farming system is based upon the principle of replacing as far as possible land and labour, which are increasingly expensive factors of production, with intensive capital investment in the form of scientific technology and the application of techniques of factory management and organisation to the production of agricultural produce from selected plants and animals. This principle of substituting one factor of production for another is one fully discussed in Chapter V, on pp. 73–87.

It is this application of principles more commonly associated with industry that has probably earned for the system its name of "factory farming", but the term has acquired emotive undertones. This is because the subjection of animals to mass production techniques has offended the susceptibilities of many people, and the term "factory farm" is consequently interpreted as implying that the livestock are treated as machines. It can fairly be stated that, as unaesthetic as some practices in this system are, much of the alleged "cruelty" lies more in the animals being deprived of natural living conditions, and that the the system is certainly less brutal than old-style cattle ranching, in which livestock in huge numbers are subjected to routine operations without resort to anaesthetic, and which has aroused little public outcry. It is reasonable to suggest that if public opinion is outraged by factory farming practice, members of the public must be prepared to forgo food at economic prices which the system has been evolved to provide.

Because of the unpopular connotations of the name, many farmers have preferred to describe themselves as 'intensive livestock producers". Like many other industries, this system is market located, and as a result differs from more conventional farming systems, the locations of which are determined by physical factors such as geology, relief, drainage, climate and soil types to a greater or lesser degree. In so far as factory farming is largely independent of the natural environment, and is instead located on the basis of economic considerations, it is of less geographical significance than the other agricultural systems. On the other hand, it may very well serve as an indicator of the ultimate character which agriculture will be forced to adopt in a "shrinking", overpopulated and increasingly technological world.

The forms that factory farming take include the "deep litter" method of pig production; battery hen production of eggs; broiler chicken production; "baby" and "barley" systems of beef fattening; fully automated dairies—the so-called "cafeteria-system" of dairy management; and the "factory" production of vegetables and flowers in a pest-free, artificial environment where heat, light and humidity are all automatically controlled. The last of these offers a possible alternative to the conventional glasshouse producer since new hazards such as "supersonic bangs" may in future threaten his livelihood.

The regional example examined in this chapter serves to demonstrate two of the above modes of production—those concerned with pigs and eggs.

Study of an Intensive Livestock Unit in Bedfordshire, U.K.

The unit is made up of parts of two once much larger farms: the Icknield Way Farm, which used to be devoted to sheep-grazing, and the Valence End Farm, which was once an arable farm given over largely to wheat (*see* Fig. 107). The business is a limited company, and leases the farm

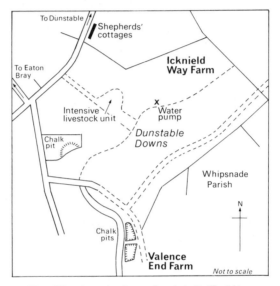

FIG. 107.—*Intensive livestock unit in Bedfordshire.*

from four lessors. The property comprises a compact holding of about 40 ha, plus a further 2.4 ha rented across the road. In point of fact the production is really centred upon an area of 4 ha of buildings and roads, and the rest of the land is only extensively cultivated.

The farm is situated on dry, undulating chalk country close to Dunstable Downs. The large quantities of water required to provide the needs of such large numbers of livestock have to be pumped from the nearby chalk hills. This water is the property of the Anglian Water Authority. The animal effluents are disposed of by draining into the chalk.

Pig production

In October 1978 there were 7,639 pigs on the farm, mainly of the Large White, Landrace and Welsh Hybrid breeds. Some of the boars were of the Welsh breed. The composition of the pig stock was as follows:

956 sows, including 683 "in-pig" and 273 "empty"
144 maiden gilts

31 working boars
8 young replacement boars
1,671 suckling pigs with their mothers
2,184 weaned pigs
2,645 pigs in the fattening sheds

The sows produce an average litter of 9.68 piglets, but the gilts or young sows slightly fewer, 8.0, giving an over-all average of 8.66. Of these young, about 8.41 per cent are born dead. The sows are artificially inseminated by skilled female operators. The breeding gilts and sows produce two litters of piglets a year and are culled at 2 years of age. All the young pigs undergoing fattening, male and female, are allowed to feed freely. At about 5 months they reach slaughter weight, approximately 55–63 kg, and about 300 fat-pigs are despatched for slaughter each week.

Egg production

In October 1978 the poultry stock consisted of:

23,000 6 to 7-week-old pullets which had been bought as day-old chicks and were being reared for stock replacement; of these a proportion were likely to be lost;

22,260 30-week-old laying pullets which came into lay at 22 weeks, and would reach their peak laying stage at 32 weeks;

21,000 68-week-old laying hens coming up to slaughter age in six weeks' time at the age of 74 weeks.

The chickens belong to the Warren Studler S.S.L. (sex-linked) hybrid which originated in France. This bird lays a brown egg, and averages 40 per cent Euro-size 1, 28 per cent size 2, 15 per cent size 3, 11 per cent "seconds" and the rest of other sizes. The average hen lays 256 eggs in its 52-week productive life, making 5,376,000 eggs during the productive life of an entire batch of 21,000 birds. During this period the food bill of the same batch will be about £106,165. When disposed of at 74 weeks the dead chicken is utilised for by-products as follows: bones for glue, feathers for meal, meat for making chicken pies and the carcases stewed for soup.

Organisation of the unit

The only crops produced on the holding include 31.5 ha of wheat. The only dressings applied are organic manure, bedding straw and wood shavings, and the yield obtained is very poor, about 6,275 kg per ha. The remainder of the farm's area, about 4.5 ha, is under grass which is cut for hay.

A large labour force is employed, including 43 full-time male workers, and 1 full-time and 6 part-time female workers. There are 3

Directors (including the Company Secretary) on the managerial staff, plus 2 male and 1 female full-time clerical workers. There are 9 maintenance workers and 1 foreman making 10 in all. On the pig husbandry side there are 17 pigmen with 3 foremen, and on the poultry side 3 poultrymen, 2 foremen and 6 part-time women employed as egg collectors (who also undertake artificial insemination in the pig section, since egg collection does not occupy all their time), making a total of 31 involved directly with the livestock. Finally there is a provender mill operated by 2 mill-workers, and there is 1 full-time lorry driver, bringing the total on the pay-roll up to 50 people.

The property and plant of the farm include two farmhouses which are occupied by the Directors, ten cottages which provide accommodation for some of the work-force, and a number of specialised livestock buildings. There is a brooder house designed to accommodate 23,000 day-old chicks, and two brooder houses each capable of accommodating 22,500 layers. On the pig-rearing side there are five farrowing houses, each with 100 farrowing pens, as there are 400 pregnant sows at any given time. There are two weaner houses each able to accommodate 1,000 weaner pigs of between 3 and 10 weeks old. There is also a fattening house to accommodate 5,000 pigs, and finally a sow house to accommodate 1,000 "in-pig" and "empty" sows and gilts and the boars. The provender mill is capable of producing about 250 tonnes of feedstuffs per week, sufficient not only for the farm's requirements, but also to take work from other farmers. Current output is about half capacity.

The sources of income of this holding are estimated as follows: pigs 65.2 per cent, eggs 32.9 per cent and other sources 1.9 per cent. These other sources include poultry by-products, manure and feedstuffs.

Part Three

WORLD AGRICULTURE AND THE WORLD ECONOMY

Chapter XVII

World Agriculture and Economic Union

Benefits and Difficulties of Free Trade

THE perfect symbiotic relationship between agricultural and industrial regions throughout the world has never been fully achieved. To begin with, complete and unrestrained free trade has been restricted by the physical barriers which occur between regions. Various economic restraints have also led to great disparities of wealth between regions, which have had the effect of making trade unprofitable. More tragic still, however, have been those restrictions upon trade which have been deliberately imposed by governments for some misguided and unfortunate ends. Customs barriers and import or export quotas have been imposed when nations have been fearful that their strategic security would be threatened if they lost their economic self-sufficiency, that is, the ability to survive blockade; or that free trade would bring about economic and, through it, political domination by some stronger nation. Furthermore, ever since the fifteenth century, "beggar-my-neighbour" policies such as "bullionism" and "mercantilism" have advocated exporting as much as possible and importing as little of neighbours' produce as could be contrived, so as to build up gold reserves at home, and develop home industries. Despite the fact that such policies provoke reprisals and inevitably lead to the ultimate impoverishment of each party concerned, they have been practised by one power or another even down to the present century, so much is one nation frequently prepared to achieve a short-term gain at the expense of another.

A policy of free trade was first advocated as the corollary of a formulated economic philosophy by Adam Smith, whose *Wealth of Nations* was first published in 1776. In this book he argued that the citizens of certain nations, because of their cultures, might be specially competent in some particular skill or craft, which gave them the ability to produce it more cheaply than their competitors. There are similar advantages accruing to regions which are naturally rich in some particular resource such as timber, a mineral or a metal, which can be procured there more readily and more cheaply than elsewhere. Such regional or national advantages may be called "comparative advantages". Smith argued that if each of several trading partners were to concentrate on producing commodities for which its territory or people enjoyed the comparative advantage, and traded freely in its surpluses for the products of other nations, all would enjoy the greatest profitability in their trade and consequently the greatest possible enhancement in their living standards. The same principle applies to domestic barter between tradesmen, so that a baker and a shoemaker are better off if they exchange the products of their skills than if they try, amateurishly, to be independent of one another and bake or cobble for themselves. A more extreme case, as cited by Paul Samuelson, is that of a secretary who types only moderately and her boss, a lawyer, who is a brilliant advocate but an equally excellent typist. It still pays the lawyer to concentrate upon law, in which he holds the greatest advantage, rather than waste his time typing where, although acknowledged to be the most skilled, his efforts can be the more readily dispensed with. There is the greatest comparative advantage in this arrangement, despite the lawyer's absolute advantage in typing.

The reasons why the irrefutable logic of free trade has not won it universal acceptance have been selfish national rivalries, short-term profiteering on the international level, and fears on the part of nations that, under free trade, those industries in which they did not possess a comparative advantage would go into decline, bringing about economic imbalance and loss of self-sufficiency. In a climate of international distrust such imbalance may be seen as placing countries in strategic disadvantage with their neighbours, and demands are therefore heard for the raising of tariff barriers and for restrictive quotas upon imports in order to restore sovereignty. For a more complete assessment of the arguments for and against free trade, the reader should consult any reputable economics textbook.

There is still another reason why universal free trade is not entirely practical. The poverty found in so many undeveloped regions leaves them few if any dependable export surpluses with which to participate in international trade. Such regional inequality of wealth is not, as might be supposed, simply the misfortune of the deprived nations themselves, obliging the more developed regions to trade exclusively among themselves. Instead it is a loss for all concerned, since, deprived of trading partners whose resources have been left untapped, none can enjoy the potential prosperity which only universal free trade would provide.

Free-trade areas and economic unions

Once the benefits of free trade had been recognised, however, there were several attempts to establish international free trade areas in the nineteenth and early twentieth centuries using one-time colonies as partners. Probably the most successful of these was the British Empire, later to become the British Commonwealth, which developed into a free-trade preference area between several large self-governing Dominions, held together by their ethnic ties and mutual advantage. In those days the political security of this "far-flung empire" depended, in what seems to us a most uncomplicated way, solely upon Britain's control of the "High Seas". Today, in an age of nuclear submarines, spying from outer space and other strategic threats which serve to make every nation insecure, "far-flung empires" are quite unthinkable, apart from ethical considerations, because of their vulnerability. Yet, at the same time, larger economic units have never been so advantageous at every level, in this age of new technologies, and so a new kind of economic empire-building is making itself manifest. Continental free-trade blocs are emerging—some by mutual decision of the members, some as a result of the economic domination of smaller states by a more powerful neighbour, yet others as a result of smaller nations negotiating membership of a larger association for economic aid and political protection.

A brilliant assessment of this situation was made as early as 1959 by J. P. Cole in his book *Geography of World Affairs* (Penguin). The following is quoted from the third edition of 1964:

> ... by the late 1960s there should be about ten large economic unions in the world, containing nearly every country in the world. If present trends continue, two will be Communist—Comecon and China, and in view of the very small trade between them and the desire of the Chinese Communists to pull themselves up by their own efforts, they should be quite separate. Two will belong to the West without any doubt, North America and West Europe. Latin America and South-East Asia with Japan will presumably be more western than neutral. The remainder—South Asia (India), the Arab World, and "Middle Africa" will be neutral. Southern Africa, with its appreciable proportion of Europeans, will not fit easily into an African bloc, while Australia and New Zealand also occupy a difficult position.

Clearly these forecasts still remain substantially valid, even though the time-scale which Cole envisaged is now seen to have been over-optimistic. The fifth edition of Cole's book published in 1979 (*see* Bibliography), is more guarded.

For the purpose of this book, in which we are concerned with the effect of such economic unions upon agriculture, it is best to devote our attention as far as possible to up-to-date facts about what is possibly the most rapidly developing and dynamic of these unions to date, rather than make any attempt to assess progress or consequences of economic union generally throughout the world. The rest of this chapter will therefore be given over to a consideration of the development of the European Economic Community generally, and its policies and achievements in respect to agriculture in particular.

The Development and Present Character of the E.E.C.

At the end of the Second World War in 1945, Western Europe was in a shambles, weak, and exhausted both economically and politically by the recent hostilities. For a while it even seemed possible that she might either succumb to direct Russian aggression, or, in seeking American aid, come under U.S. economic domination. The U.S.

government of the day recognised, however, that the interests of herself and the entire world were best served by the restoration of a strong Europe. It was proposed to the leaders of the war-torn nations that U.S. monetary aid would be made available to them only on condition that a co-ordinated plan for post-war reconstruction and future economic co-operation was put forward.

Already the International Bank for Reconstruction and Development and the International Monetary Fund had been established. Now a corresponding Organisation for European Economic Co-operation was set up to administer Marshall Aid from the U.S.A. to Europe.

In 1952, following the Treaty of Paris, the European Coal and Steel Community (E.C.S.C.) came into existence between the following six signatories: France, West Germany, Italy, Belgium, Luxembourg and the Netherlands. In 1957 the same "Six", as they had then become known, formulated and signed the two Treaties of Rome, which established two further communities: EURATOM, the European Atomic Energy Community, and the E.E.C., with which, in considering the agriculture of Western Europe under economic union, we will be primarily concerned here. As from January 1958, when the E.E.C. came into operation, the following conditions were made binding upon the "Six".

1. Customs duties and import and export quota restrictions between member countries were to be gradually lowered and abolished. This was finally achieved, for all but certain types of agricultural produce, by July 1968.
2. A common external customs tariff and a common commercial policy towards non-members were to be established and maintained.
3. Freedom of movement for various categories of labour, services and capital within the market was to be established and maintained.
4. A common set of policies for agriculture, transport and trade were to be established and maintained.
5. Associations between the Community and countries outside the Community were to be developed for their mutual advantage. Provided such countries were in Europe and met certain conditions, such as having a democratic form of government and having reached a certain level of economic development, they could in due course be considered for full membership.

In recognition of the fact that the ultimate purpose of the Community was no less than political union, the Council of Ministers, the Commission, the Court of Justice and the European Parliament were set up, with legislative, executive, judiciary and consultative functions respectively. In 1979 the European Parliament was constituted with directly elected members, and it is intended that its legislative powers should be extended in due course.

In 1971 plans were drawn up to admit the U.K., Ireland, Denmark and Norway into the Community. Norway subsequently did not ratify the Treaty of Accession and therefore did not become a member, but the others joined to form an enlarged community of "Nine" with effect from January 1973. In addition to neutral associated countries such as Switzerland and Sweden, the present list of candidates for membership includes Greece (whose membership is due for ratification in 1981), Spain, Malta, Israel and Turkey. Possibly such countries as Finland, Yugoslavia and Egypt may subsequently be considered for admission, and should the U.S.A. ever negotiate entry, a North Atlantic Community could emerge. Trade links have also been forged with one-time colonial associates of the members, such as some North African countries, e.g. Morocco and Tunisia, and New Zealand, once under French and British rule respectively.

A strong European Parliament and a stable and accepted common European currency both need to be achieved, preliminary to establishing closer federal ties between the member states.

Possibly the most difficult sector of the E.E.C. for which to formulate acceptable policy guide-lines and regulations has been agriculture, to which we now turn.

Agricultural Policies and Achievements in the E.E.C.

The Treaty of Rome established guide-lines for a Common Agricultural Policy (C.A.P.). The aims of the Policy would be to:

1. increase agricultural productivity;
2. ensure a standard of living for those engaged in agriculture comparable with that of other workers;
3. stabilise markets for agricultural produce;
4. ensure adequate supplies at reasonable prices to the consumer.

These aims have to be achieved within the framework of free trade between member countries in farm produce, joint financing of the system by the members, and as far as possible common levels of support for producers and consumers, and a common policy towards non-member countries.

It has been a constant concern of the Community to reconcile the interests of member countries seeking special consideration for a particular sector amongst their agricultural producers or for their consumers penalised by a particular feature of supply, and the interests of the Community as a whole seeking to establish a unified agricultural system beneficial and acceptable to all.

The achievement of free trade in farm produce is intended to bring about common price levels for a wide range of agricultural products throughout the trade area. Such a policy is, of course, inevitably going to penalise "expensive" producers, who, if they cannot pass on their high production costs to the consumer, are no longer as profitable as the efficient producers, or, if they continue to charge high prices, cease to be competitive. They are obliged to find some way of reducing production costs, switch to some mode of production in which they can enjoy a relative advantage, or else go out of business. It is E.E.C. policy to allow a transitional period so that producers can adjust themselves to the freer competition, and to make funds available for "agricultural guidance", explained subsequently, to help in "structural reform", for example, land-holding consolidation, or assistance to people wishing to leave farming, or else to facilitate "long-term projects" such as reafforestation, drainage, or the redeployment of productive capacity from products in constant surplus to those in which a deficit persists.

In these policies the Community is only helping to accelerate much-needed rationalisation and trends towards ever-increasing size for productive units, necessitated by world economic forces, e.g. a growing need to produce more and comparatively cheaper food on a declining area of available farm-land, and made practical by improvements in technology. Policies of protectionism do little more than "delay the evil day", and by preserving anachronistic survivals can only serve to handicap national economies which may be impoverished by the struggle to assume an ever-mounting burden of support for inefficiency.

Already the E.E.C. members have experienced a substantial reduction in their agricultural work-forces, as shown in Table 20.

TABLE 20

Reduction of agricultural work-forces in E.E.C. countries

Date of E.E.C. membership	Country	Percentage of working population engaged in agriculture	
		1969	1975
1958	France	14.6	11.3
	West Germany	9.4	7.3
	Italy	20.7	15.8
	Netherlands	6.6	6.6
	Belgium	5.3	3.6
	Luxembourg	16.0	6.2
1973	U.K.	2.75	2.7
	Ireland	27.2	24.3
	Denmark	13.0	9.8

Achievements of the agricultural guidance policy

Since 1958 the number employed in the agricultural sector of the countries which make up the "Nine" has been reduced from 19 million to 8 million, with the biggest decline being registered amongst farm labourers. In the "Six" the average annual reduction has been about 2.7 per cent, but this has concealed such variations as a 10 per cent fall in France and Italy, where previously agriculture was under-mechanised and overmanned, and negligible reductions in more-efficient Belgium and the Netherlands. Since joining the E.E.C., the U.K., Ireland and Denmark have only displayed a 2.1 per cent annual reduction, a fact not so surprising since they had already experienced reductions previously, particularly the U.K. and Denmark. Indeed, the agricultural sector of the U.K.'s working population has been declining for the past two centuries. Hill regions and regions with a lack of alternative employment are among those that require special aid from the E.E.C. in order to make the necessary transitional changes, e.g. the U.K.'s north-west uplands, Ireland, southern Italy, and Mediterranean France (*see* p. 284).

Another move towards greater agricultural efficiency has been the restructuring of land-holdings. Even between 1973 and 1975—the first two years after the U.K., Ireland and Denmark joined—the average farm size in the E.E.C. rose from 16 ha to 24 ha. In order to achieve larger, more efficient units, these amalgamations are encouraged in two ways. First, the E.E.C. pays 25 per cent grants towards the cost of such amalgamations, or even 65 per cent in Ireland and parts of Italy. Secondly, there are 25 per cent grants for approved five-year farm development programmes designed to raise farm income.

The increased efficiency which arises from larger farm size is essential because industry and housing keep encroaching upon the total agricultural land available. In the period 1968–74 in the original "Six" alone some 1.2 million ha of good agricultural land were lost in these ways.

Farmers seeking development grants have to prove their competence, be willing to co-operate in compiling a satisfactory development programme, and be answerable for keeping a record of progress made. The grants are only intended for farmers dependent upon their farm income, and for farms potentially capable of development. Such plans can include land consolidation, drainage and irrigation schemes, development of machinery syndicates, and desirable switches of production, e.g. from milk to beef. Purchases of land, pigs or poultry are not eligible for grants.

Grant-aided full-time training is available for agricultural advisers, farmers and farm workers, and such workers are encouraged to work towards advanced professional qualifications. There are also in-service schemes designed to help farmers keep abreast with the rapid developments in agricultural technology.

To encourage the release of land for amalgamations for the establishment of larger, more economic units, farmers of between 55 and 65 years are encouraged either to retire or to find alternative work, for which they are offered an annuity or lump-sum payment. Farmers and farm workers of 60–65 in most countries, and of 55–65 in Ireland and southern Italy, are given help to retire, in order to reduce the percentage of the work-force dependent upon agriculture. The land thus released can either be sold or leased for a minimum of twelve years to farmers who wish to develop their holdings, or less-desirable land can be diverted to non-agricultural uses such as public utilities, reafforestation or recreation (*see also* p. 120).

The importance of such Community aids to agriculture during the difficult period of transition towards fuller economic integration is heightened by the various ways in which the member countries are dependent upon agriculture. Thus for some countries it is a major part of national gross income: in Ireland 19 per cent, Denmark 9 per cent, and Italy, France and the Netherlands more than 5 per cent. Only for Luxembourg, Germany, Belgium and the U.K. does agriculture constitute less than 4 per cent of national product. Yet even these countries may have other reasons for valuing the contribution of agriculture to their economies. For Ireland agriculture is the source of 50 per cent of her exports, for Denmark 30 per cent, for the Netherlands 25 per cent and for France 20 per cent. Major food-importers are concerned to keep some control on their balance of payments. These include Germany, Italy and the U.K., for each of which food constitutes 20–25 per cent of their total imports, and where any increase is regarded as undesirable in view of the insecurity of the world market situation.

The "joint schemes" for agricultural guidance are financed by the individual member countries out of their own budgets, and refunded to them in part out of the Community budget.

Aid to farmers in regions of difficulty

For social reasons, aid to hill regions or to regions which lack alternative sources of employment is also an approved part of Community policy. Such regions include the Scottish Highlands, Lake District, Pennines, and North and central Wales in the U.K., the Massif Centrale in France, and southern Italy; indeed, such areas make up 25 per cent of Community farm-lands. The measures may either take the form of direct payments per head of cattle or other livestock, or per unit of land on the holding; or involve interest rebates or deferred payments on grants of capital. All the measures are designed to minimise economic hardship in the poorer agricultural communities and so prevent or retard rural depopulation.

Guarantees to stabilise agricultural supply and demand

So far we have considered how the European Agricultural Guidance and Guarantee Fund (F.E.O.G.A.) fulfils the "guidance" function. By far the largest part of the fund, however, indeed 65 per cent of the Community budget, is spent on guarantees to the farmer and subsidies to the consumer.

Policies have been created to improve marketing, not only in order to stabilise farm income, but also to improve food supplies and stabilise food prices to the consumer. Such a situation of stable supply and demand first requires the stability of the national currencies within the Community. The long-term aim has always been to establish a single stable currency throughout the Community to provide common support prices for producers and ensure uniform prices for agricultural, as for all, commodities, throughout the Community. This was to be achieved by giving rein to natural equilibrium adjustments which take place under a free-trade policy. These ideals remain unfulfilled, however, because of conflicting patterns of economic development and the difficulties of adjustment from sovereign state to federal state experienced by various countries within the Community.

Since national currencies have appreciated and depreciated in relation to one another or been allowed to "float" against one another, it has been quite impossible to apply common price levels throughout the Community. To compensate for this, farm support prices are calculated by special exchange rates called "green rates" which have been negotiated to meet the needs of particular countries instead of basing market and consumer prices upon current rates of exchange. The Community makes up the differences between the "green" currency and the true national currency. When the "green currencies" are either lower or higher than the real currencies, this can be made up for by "monetary compensatory amounts" taking the form of either "border levies" (charges) or "subsidies" (payments) as food crosses a national boundary. Food from outside a country must also pay an adjusting higher or lower tariff charge. The same device also prevents a country which is importing subsidised foodstuffs from re-exporting them at a profit, since they then carry a levy. These "green rates" cost the Community a great deal, and there has been pressure within the Community for the "green rates" to be kept closer to real currency values. There have also been efforts to reach the ultimate aim of a unified European currency, which would resolve the problem of negotiating unsatisfactory compensatory devices like the "green rates" for food.

After food has been produced, the costs of processing, distribution and marketing may double the final price at which it reaches the consumer. This final price, nevertheless, does reflect the prices fixed by the Community.

The E.E.C. produce markets differ in the details of their operation for each of a number of main food commodity groups, as illustrated below. At present, E.E.C. regulations cover 96 per cent of the Community's farm produce, which includes tobacco, hops and wine, in addition to the following.

1. Cereals.

(*a*) A "target price" is set by the Community for each type of cereal at the price the grain would fetch on the open market at Duisberg, in the Ruhr Valley, where grain is in shortest supply. This is the price of grain delivered to the merchant.

(*b*) A "threshold price" for grain at ports of entry is calculated so that at this lower price it would reach target price at Duisberg. Grain reaching E.E.C. ports at below threshold prices is liable to a levy to bring it up to threshold price, thus preventing outside producers under-cutting those within the Community.

(*c*) There is also an "intervention price". So long as there is no surplus production on the home market, the threshold price operates as an effective import control, and serves to stabilise the home market prices. If, however, there is a production surplus at home, National Intervention Agencies (eleven in each of the U.K., Germany, France and Italy, eight in Denmark, five in Ireland, and two each in Belgium and the Netherlands)—will purchase grain offered to them by traders at between 12 and 20 per cent below target price. This intervention buying will help to "firm up" the market price when it begins to sag. The grain purchased may be either sold back to a dealer when the price has recovered, or sold abroad, in which case it will be subject to an export levy if prices are higher, or an export subsidy if world prices are lower. Alternatively, the grain can be processed into another commodity such as malt or starch, and the revenue from its sale repaid to the Community.

Producers of "durum wheat", the hard wheat used in pastas, are paid special deficiency payments to encourage increased production in order to reduce Community imports from suppliers outside the Community.

2. Sugar. Under the Lomé Convention (explained subsequently, *see* p. 297) the Community is committed to accept 1.3 million tonnes of cane-sugar from such developing countries and regions as Fiji, Mauritius, the Caribbean and India whose economies are dependent upon it. Such producers enjoy a guaranteed price linked to the Community's internal support prices. However, the bulk of Community sugar is from domestic beet production and from cane-sugar produced in France's overseas territories. Each year a target price is set for the main sugar-producing regions, supported by an intervention price at which the Community agents will step in and buy sugar which has been made available to E.E.C. sugar-deficit regions. This full intervention price is only allowed for a basic quantity of 9,136,000 tonnes from the E.E.C. producers. The "basic quantity" is then split into basic quantities allowed for each member state to produce, and this is again split into production quotas for each sugar factory in that state's territories. Above this "basic quantity", a "maximum quantity" is ascertained each year; in 1977–8 it was fixed at 35 per cent above "basic". In this additional quota a production levy is imposed on sugar (30 per cent in 1977–8) which effectively reduces the value of the intervention price as a means of support.

Sugar produced in excess of this "maximum quantity" receives no support whatsoever. Factories impose prices upon farmers which are directly in proportion to these levels of support. Those farmers supplying beet for the "A" or basic sugar quota receive a higher price than those supplying beet for the "B" or maximum sugar quota, and those supplying sugar in excess of the maximum quota receive no minimum price protection. Free market imports of sugar from outside the Community are subject to payment of import levies to bring their prices up to fixed threshold prices, thereby reducing any advantage they might have had over Community producers. When world prices are low, compensatory prices are paid to exporters to help lift E.E.C. prices, whereas when world prices are high, export levies may be imposed to prevent a rush to export.

During an emergency such as the Community sugar shortage of 1974–5, special purchases can be made outside the Community in face of world competition for special, heavily subsidised sale to the consumer.

3. Milk and milk products. Under regulations that came into operation in 1968, a "common target price" is established for milk based upon what producers should receive on delivery of milk to dairy, assuming access to markets in and out of the Community. Of the milk produced, 75 per cent is processed, and only 25 per cent sold fresh. The policy provides for "intervention prices" for butter and skimmed milk generally, and cheese from Italy in particular, in order to sustain the target price. "Threshold prices" are also assessed for twelve "pilot products", and importers are required to pay a levy to make up lower priced imports to the Community price, in order to prevent milk imports depressing E.E.C. prices. Threshold prices for other dairy products are based upon these pilot products.

In order to reduce overproduction of milk in the E.E.C. there are various schemes to help farmers switch to other production. There are also schemes to boost demand, such as consumer subsidies for butter, provision of school milk, and assistance to consumers of skimmed milk. Substantial stocks of dried skimmed milk are regularly sent to developing countries each year. New Zealand producers, so long dependent upon the U.K. market, enjoy a guaranteed price for an export quota of their butter to the E.E.C., although no such agreement has been achieved so far for their cheese.

4. Beef and veal. To restrain the trade fluctuations which harass beef and veal producers, heavy but flexible tariffs are imposed upon imports of these products and of live cattle. Thus, in addition to a 16 per cent tariff on live cattle and 20–26 per cent tariff on beef imports, there are also flexible levies payable by importers, subject to constant review in the light of market conditions. In an emergency, first these, then the tariffs may be temporarily suspended.

5. Pigmeat. Although not subject to regular support, the Community occasionally intervenes when prices fall 8–15 per cent below a "basic price" assessed by the Management Committee for pigs. The same committee also establishes a "sluice-gate price" for pig and pig-meat importations. A levy is imposed to bring imported goods up to this price, and in addition, a variable levy is chargeable, which is made up of two elements:

- (a) the difference between pig-food prices in and out of the E.E.C.;
- (b) 7 per cent of sluice-gate price, giving the domestic producer a substantial advantage over outside competitors.

Exporters of processed meats made in the Community also enjoy a "restitution payment", a form of subsidy to make them competitive.

6. Eggs and poultry. Harmonisation of cereal prices between the "Six" has made possible the creation of a marketing policy since July 1967 for poultry and eggs, which depend so heavily on cereal-based feed-stuffs. Imports of eggs and poultry from outside are regulated by levies, very like those imposed on pig-meat imports, i.e. a supplementary levy is imposed to bring the prices up to a sluice-gate price based on world production costs as assessed by a Management Committee for eggs and poultry, plus a 7 per cent variable levy based upon the higher cost of feedstuffs in the E.E.C. plus an element of preference.

7. Fruit and vegetables. In this field the regulations are far less rigid, a policy dictated by the more local and seasonal character of production, and the perishability of the products. A "reference price" is established on the basis of costs of production and marketing. Imported goods are required to pay (a) a customs duty of 10–21 per cent for vegetables, and (b) a countervailing duty, when the price of imports falls below the reference price on more than two successive days. For home-produced tomatoes, cauliflowers, grapes, peaches, apples and pears the Council of Ministers fixes an annual "basic price", and member states fix a "buying-in" price at 40–70 per cent of this basic price. Like the intervention price established for other commodities, the buying-in price is the point at which governments step in to support the market price.

Strict grading standards have been established for fruit and vegetables, and domestic intervention is limited to "buying-in" at the level of Grade II goods, thereby giving an incentive to producers to sell their best-quality produce further afield in the market. Producer organisations are provided with grant aid for their establishment, but are subject to disciplines for inefficiency.

8. Potatoes, mutton, lamb and wool. Although there are plans to manage the market for these commodities under the Common Agricultural Policy, existing market controls are confined to those set up by national governments. Import duties on mutton and lamb stand at 20 per cent, and on live sheep at 15 per cent. Wool, however, as an industrial product, enjoys neither import duty nor Community support, although some marketing operations are supported by member states.

The Present State of Agriculture within the "Nine"

The members of the Community may be ranked in the order of their agricultural production as follows: France, Italy, West Germany, the U.K., the Netherlands, Denmark, Ireland, Belgium and Luxembourg. The following pages examine the present state of agriculture within the individual countries.

France

France is the dominant agricultural producer of the E.E.C., contributing 50 per cent of its wheat, 60 per cent of its maize, 28 per cent of its barley, 31 per cent of its beef, 31 per cent of its milk and 33 per cent of its refined sugar. Agricultural production contributes 5.7 per cent of gross national product.

The agricultural labour force constitutes 11.3 per cent of the total labour force. Only a declining 13 per cent of the agricultural labour force is made up of labourers, and of the remainder of this labour force, 30 per cent is over 60 years of age. About 53 per cent of the farms are owner-occupied.

As regards size of farms, 30 per cent of the holdings are between 20 and 50 ha, while another 20 per cent are less than 5 ha. Out of the 59 per cent of total land area given over to agriculture, 52.7 per cent is arable, and 41.4 per cent is in permanent grass. An alarming 75,000 ha of land are lost to farming annually as a result of industrial and urban development.

To help small farmers and market gardeners remain competitive, co-operative marketing has been extensively developed, so that today 90 per cent of all cauliflowers and early potatoes, 70 per cent of cereals, 60 per cent of wine, 52 per cent of apples, pears and tomatoes, and 40 per cent of milk are marketed co-operatively. France, however, has a deep-set problem of regional disparities in land-holdings and income.

Italy

With her large population, Italy is a net food importer, but she is nevertheless a major agricultural producer, and some 8.2 per cent of her national income is derived from farming. The main products are fresh fruit and vegetables, wine, maize and olive oil. Italy produces 47 per cent of the Community's wine and 33 per cent of its maize. Some 15.8 per cent of Italy's total labour force is engaged in agriculture. About 75 per cent of the farms are owner-occupied, but most of the farmowners are over 50 years of age and poorly educated. A large proportion of the agricultural labour force—44 per cent—is made up of farm labourers.

Farm-holdings are very small—68.4 per cent are less than 5 ha and only 2 per cent are over 50 ha. About 58 per cent of the total land area is given over to agriculture, of which 52.8 per cent is arable and 29.7 per cent in permanent grass, while 17 per cent is used for permanent crops such as olives and vines.

It has been difficult to promote co-operative marketing in such a country of small backward farms, but today some 50 per cent of the farmers are members of *Consorzi Agrari*. Despite rising production costs and rampant inflation, Italian farmers have enjoyed improvements in their living standards. There are plans for new irrigation projects to revolutionise agriculture in the Mezzogiorno region of southern Italy.

West Germany

West Germany has a large industrial population, and is therefore a net food importer, deriving only 2.8 per cent of her gross national income from agriculture. Nevertheless, she contributes 30 per cent of the Community's pigs, 18 per cent of its beef, 30 per cent of its potatoes and 80 per cent of its rye and oats.

The agricultural labour force constitutes 7.3 per cent of the total labour force. About 71 per cent of the farms are owner-occupied, but the farm population is ageing, with 30 per cent of the farmers over 55 years of age. Farm labourers constitute only 8 per cent of farm labour, and so, as the work-force ages and dwindles, increased mechanical and technical inputs are being applied. There is also an increased tendency towards part-time farming.

The average size of farm-holdings is low, with 34 per cent below 5 ha and fewer than 3 per cent above 50 ha. Out of the 53.7 per cent of total land area devoted to agriculture, 56.6 per cent is devoted to arable farming, 39.4 per cent to permanent grass, 1.5 per cent to permanent crops and 2.5 per cent to kitchen gardens.

Despite a general lack of incentive in a food-importing country, the government encourages co-operative marketing for many products.

Today 70 per cent of farms belong to co-operatives, which market 80 per cent of milk output and 50 per cent of cereals. In addition, nearly all the sugar-beet and peas are produced under contract.

Since Germany had for a long time been a high price area, the establishment of the common agricultural prices in the 1960s actually reduced prices to the farmer. Nevertheless, German farmers have subsequently benefited not only from the "Green Deutschmark" which maintains agricultural prices at a level higher than the national currency, but also from the country's relatively low rate of inflation.

The United Kingdom

The United Kingdom, by virtue of her large population, is the major food-importing country of the Community, importing 45 per cent of her requirement. Yet at the same time she is a major producer, and her land is intensively and efficiently farmed. Two-thirds of her agricultural output is in livestock and animal products, particularly milk and milk products, fat cattle and calves, cereals and potatoes. She contributes 44 per cent of the Community's sheep, 26 per cent of its barley and 22 per cent of its eggs. Because she is heavily industrialised, however, agriculture constitutes only 2.2 per cent of gross national income.

Only 2.7 per cent of the U.K.'s total labour force is engaged in agriculture. About 60 per cent of the farms are owner-occupied, and the U.K. continues to employ a large proportion of hired labour—39 per cent. The labour force is comparatively young.

The average farm-holding is large, three times that of France and ten times that of Italy, thanks to centuries and more of consolidation and enlargement. Nevertheless, 14.5 per cent of holdings are less than 5 ha, though these are mainly in marginal regions such as the Scottish Highlands. Large disparities of farm size and income persist throughout the U.K. The country has the largest area in agricultural use anywhere in the E.E.C., 76.4 per cent. Against this, much of this land is marginal, with 62.5 per cent in permanent pasture and only 37 per cent arable. Much of the former is fit only for sheep.

When the U.K. joined the Community in 1973 her whole system of subsidies and markets had to be radically changed to conform to those of the E.E.C. For example, whereas Community farmers sought credit from government agencies, British farmers were accustomed to applying to commercial banks and agricultural merchants, and receiving tax concessions from the government for capital investments and building improvements.

Since the Agriculture Act 1947, British farmers have received direct subsidies for capital improvements, and have enjoyed guaranteed prices for their produce. If imported goods have depressed farm producers' prices, then the deficiency payments system has made them up to guaranteed price levels. All this has now been swept away. Instead the

British farmer finds himself amply protected from world competitors, but wide open to competition from farmers within the Community— as well as free to compete for their traditional markets. Bereft of his production subsidies, he is permitted instead to recoup all his production costs and secure his profit margins direct from the consumer. He enjoys a level of support from the Community intervention prices, designed to sustain the home market when it is depressed by overproduction. The need for greater competitiveness has encouraged farmers to enlarge their holdings still further, step up their investment and develop larger and more consolidated marketing organisations, to achieve greater economies of scale.

Beef, milk, cereal and sugar producers stand to gain most under the E.E.C. system. As regards beef, the new British Intervention Board for agricultural produce is able to draw upon the experience of the Meat and Livestock Commission, which helps to supervise the price intervention system for the beef market. The British support buying price stands well above the old guaranteed price level. British farmers are favourably placed to export beef to a large and expanding consumer market, and are accordingly exporting beef and live cattle to France and Belgium. The main problem for British producers is the rising cost of cereal-based fodders, which is prompting substitution with soyabeans or tapioca, and switches to less-intensive grazing systems.

The milk producer also benefits from the more comprehensive new system. British dairy farmers had previously benefited from the Milk Marketing Board through which all milk was marketed, but the guaranteed price was based solely on liquid milk and there was no support for any surplus sold as cheese or butter, so that sale of these depressed the dairyman's over-all income. Under the E.E.C. system there is support for butter and skimmed milk as well, which indirectly, if not directly, also means support for cheese production.

Cereal producers have the benefit of a number of Intervention Centres, set up in February 1973, at which the co-operatives or corn merchants can "sell to intervention", i.e. sell their produce at an agreed support price, if and when the price on the open market is depressed to what for them is a dangerous level (*see* p. 286). So far, since the Centres were set up, world prices have never fallen to the point where such intervention has been necessary. The higher prices now required for cereals have put additional pressure on the producers of beef, pig-meat, poultry and eggs, who have depended upon cereal-based feedstuffs. This is particularly serious for the pig-meat producers and processors who do not enjoy similar support. Cereal producers often combine cereals with sugar-beet, which also enjoys E.E.C. support, and with potatoes, which have the support of the Potato Marketing Board in the U.K., and which can be sold either for human consumption or for fodder.

Not only do fruit and vegetable producers, and those of poultry and eggs, enjoy a level of support from the Community, but also their comparatively large scale of production and marketing organisation in the U.K. equips them adequately for competition with their continental counterparts.

So far, however, producers of pig-meat and those engaged in processing pork products have no such support, and must rely on their expertise in order to compete with continental rivals in Denmark and the Netherlands, in face of the rising cost of feedstuffs.

Similarly, the British hill farmers, heavily dependent upon lamb, mutton and wool production, none of which are so far supported, would be severely disadvantaged but for the aid they receive from the Community as part of the package deal for "farming in difficult areas", which will enable some farms at least to be restructured and achieve viability.

From the U.K. point of view, it has been the severe inflation which has eaten up the increased market prices, while the "Green Pound" has served only to subsidise the consumer at the expense of the profit margins of the producer, which have been held low.

The Netherlands

The Netherlands has always been a small but highly efficient agricultural producer, and agriculture constitutes 5.4 per cent of gross national income. Agricultural products also make up 25 per cent of total exports. Major products include beef and dairy cattle, pigs, milk, wheat and horticultural produce.

Agricultural employment constitutes 6.6 per cent of the total employment in the country. Half the farms are owner-occupied, although reclaimed lands are usually rented out by the government. Only 13 per cent of the agricultural labour force are labourers.

With her large population density—332 per km^2—and the resultant land shortage which has encouraged large-scale reclamation, it is hardly surprising that farm sizes are low even by E.E.C. standards, and despite recent progress in amalgamation. Thus 30.6 per cent of farms are between 10 and 20 ha. Of her total land area, 58.1 per cent is devoted to farming, with 59.3 per cent of this in permanent grass which carries twice the livestock densities of the best French pasture-lands—a fact made possible by twice the density of applications of nitrogenous fertilisers of anywhere in the Community.

Dutch agricultural marketing is highly developed, with co-operatives handling 95 per cent of horticultural produce and 85 per cent of pig-meat. Dutch farmers are anxious to retain this lead. The government of the Netherlands provides loans to farmers for up to twenty-five years, and agricultural advisors are more readily available there than anywhere in the Community.

Denmark

Despite her small size, Denmark is an important exporting country, and indeed, her agricultural exports constitute 30 per cent of total exports and 8 per cent of gross national income. Pigs and beef represent 90 per cent of her farm output, of which 70 per cent is exported—40 per cent to the U.K. The main products include pigs (16 per cent of all farmers specialise in intensive pig production), beef, dairy produce and barley. About 7 per cent of her entire working population is engaged in agriculture, and practically all the farms are owner-occupied.

Family labour is usually sufficient—indeed with intensive application of machinery an increasing number of farmers also work in industry. Consequently there are only 12 per cent of farm labourers amongst the farm work-force. The age structure of Denmark's agricultural labour force is very well balanced.

One-third of the farms are between 20 and 50 ha, and only 11.9 per cent are less than 5 ha. The Danish government regards holdings of about 100 ha as approaching optimum size, beyond which "diminishing returns" are likely to set in. Of the 68 per cent of land area given over to agriculture, 90 per cent is arable, which is increasing at the expense of the permanent pasture. Cereal yields are high, indeed Danish wheat yields are the highest in the E.E.C. The Danish co-operatives are large and efficient, and have in recent years undergone much amalgamation, restructuring and vertical integration, that is, nearer to a producer-to-consumer situation (*see* Chapter XV).

In Denmark, advisory services must be paid for, and the farmer, although well educated in general, is not obliged to have formal agricultural training.

Denmark has benefited from Community price stability, although she has had to face intense Dutch competition for her traditional markets. Denmark's agricultural productivity, which has always been one of the highest in the world in proportion to her size, has not increased as much since joining as in the other eight E.E.C. countries.

Ireland

Ireland is a small poor country, in which agricultural produce constitutes 18 per cent of gross national income and makes up 50 per cent of her overseas trade, mainly with the U.K. Ireland's chief products include dairy and livestock produce. She exports 85 per cent of the cheese, 66 per cent of the butter and 85 per cent of the beef she produces.

Practically 25 per cent of her total working population depends upon agriculture for a livelihood, but her population density, 44 per km², is the lowest in the Community. Most farmers are owner-occupiers, and all but 14 per cent of the farm labourers are members of the family they work for. A quarter of all the farmers are over 65 years of age. Very little fertiliser or machinery is used, and yields are accordingly very low.

Some 30.5 per cent of all farms are in the 10–20 ha size range, but the government is seeking to encourage farm amalgamations with an incentive bonus scheme.

Of the 60 per cent of total land area given over to agriculture, 75.2 per cent is in permanent grass, for which Ireland, with its damp, cool climate, is best suited. Improvements are financed by the Agricultural Credit Corporation, originally set up with government support but now self-financing. Government grants to agriculture, including one to cover small farmers' rates, are generous. Since joining the E.E.C., Irish farmers have benefited from higher prices, farm improvement grants and improved marketing facilities, and already all peas and sugar-beet are produced under contract.

In the years 1970–5 Ireland's agricultural productivity grew faster than that of any country in the Community other than the Netherlands, despite the fact that farm costs grew four times faster in Ireland than in the Netherlands.

Belgium

Thanks to her agricultural efficiency, Belgium is a net food exporter, despite the fact that her industries are also highly developed, as evidenced by the fact that agriculture constitutes only 2.7 per cent of gross national income. Pigs (27 per cent of output), beef cattle, poultry and milk are her chief products.

Only 3.6 per cent of Belgium's working population is engaged in agriculture, and of these only 5 per cent are agricultural labourers. Some 70 per cent of all farms are tenanted, and half the farmers are over 50 years of age, making this farm population structure one of the most precariously balanced in the E.E.C.

With a population density of 321 per km^2 the holdings tend to be small—about 30 per cent being less than 5 ha in extent. Of the total land area, 60 per cent is given over to agriculture, of which 50 per cent is arable, 47.2 per cent permanent grass and 2 per cent allotments and kitchen gardens.

Credit for farm building or mechanisation is provided by the Agricultural Co-operation Bank. Marketing is also well developed, with 100 per cent of peas and sugar and 90–95 per cent of poultry being produced under contract. Since joining the E.E.C., milk producers have experienced a fall, but pig producers a rise, in income. In general, market prices for agricultural produce have fallen, but farmers have earned sufficient to offset both this and the rising cost of living.

Luxembourg

Despite her rugged topography, Luxembourg practically achieves agricultural self-sufficiency with 90 per cent of her output in animal products, the most important being beef, milk, and pig-meat. Barley and oats are the chief crops. Agriculture contributes 3.3 per cent of gross national income.

Agriculture provides employment for 6 per cent of the total labour force, but people are moving over into the industrial sector. Only 5 per cent of the total agricultural labour force are farm labourers, and farmers have therefore invested heavily in machinery and fertilisers. At present 60 per cent of the farms are owner-occupied, but this number is declining.

Holdings are large, with 41 per cent of farms being between 20 and 50 ha. Amalgamations of holdings have gone forward rapidly since joining the E.E.C. Of the 51 per cent of total land area used for agricultural purposes, nearly 54 per cent is in permanent grass for milk and beef production, most of the remainder being used for arable production, although some is used for wine.

Some 90 per cent of all farmers are members of co-operatives, which handle 90 per cent of milk output and 70 per cent of cereals.

In general, Luxembourg, like Belgium, has enjoyed a stable currency, and agricultural prices have changed very little under the E.E.C.

E.E.C. Aid to Developing Countries

Many so-called "Third World" countries have a largely agrarian economy and, particularly when the situation is aggravated by over-population, they may be held in a "vicious circle of underdevelopment". More serious still, living standards in such circumstances may actually decline, as a result of rising population, and because subsistence farming systems cannot, of themselves, generate the investment capital necessary if economic development is to go forward. In other words, modern commercial agriculture can only be established in such regions if industries are first developed in order to provide agriculture with technological support and the necessary markets for its produce.

Earlier in this chapter it has been observed that even the technically advanced countries are finding membership of larger supra-national free-trade blocs essential if they are to keep abreast of modern economic developments. It is therefore not surprising that underdeveloped regions are obliged to look to one of the major continental powers, the U.S.A., the U.S.S.R. or China, or to one of the new free-trade blocs, as being the most probable sources of development aid.

Under the Rome Treaty, member states of the E.E.C. agreed "to associate with the Community the non-European countries and territories which have special relations with Belgium, France, Italy and the

Netherlands". Such associations have from the beginning included agreements on trade, investment and technical aid, but over the years programmes have been extended to include territories associated with recently joined members, and over much larger areas of the world. Thus whereas the Yaoundé Agreements of 1963 and 1971 involved aid eventually to nineteen African states, the first Lomé Convention of 1975 has been extended to associates in Africa and the Caribbean and Pacific Regions (the "A.C.P." countries). From an original forty-six signatories, the number has grown to fifty-three. The Convention provides for:

1. duty-free access to the Community for 99 per cent of all agricultural exports (only products regulated under the C.A.P. are excluded) and all industrial exports of the A.C.P. countries, with guaranteed prices for certain raw materials (including twelve products upon which A.C.P. countries depend, such as ground-nuts, cotton and hides), to provide an assured revenue to these countries in the event of world trade recession or production problems such as drought;
2. agreement to buy 1.2 million tonnes of white sugar each year from A.C.P. countries; rum also to be duty free;
3. financial and technical aid with priority given to rural development (27 per cent of the total) and social facilities (15.5 per cent).

Plans are already formulated for a second Lomé Convention operable from 1980 onwards when the present agreements expire.

Part Four

FIELDWORK

Chapter XVIII

Fieldwork and the Use of Field Data

The Value and Purpose of Fieldwork

IN educational terms, there are two main justifications for fieldwork: first, that "the ground . . . is the primary document", i.e. it is the ultimate source of new geographical discovery; and secondly, that fieldwork is an essential part of all students' education, the only way for them to acquire personal experience of geographical phenomena about which they have been taught, and also develop the practical methodology which equips them, in turn, to undertake fieldwork for the sake of subsequent discovery. In writing this book, the second, educative, purpose stands uppermost.

The presentation in this book of many sample studies will inevitably prompt the criticism that such studies place too great an emphasis on the particular and the atypical. Such criticism may too often be used to justify what would seem to be an equal error. In order to correct possible inaccuracies in conclusions derived from too limited sampling, much wider random sampling is undertaken and efforts are made to standardise the results. In recent years the daunting task has been made more practical by use of the computer, into which the data acquired can be fed, and subsequently evaluated, in a minimum time. The value of such generalisation is without doubt considerable. What is in question is a growing tendency to subject students to such abstracts without their first acquiring personal knowledge and insight of the concrete realities upon which the abstracts were based. That is why the samples in this book are presented—anticipating the criticism, but in the conviction that the student will find plenty of more generalised studies in the geography of agriculture and will appreciate them more if they either have read this book first, or have it available for consultation. Ideally students, where time and opportunity permit, should also undertake their own field studies, to provide themselves with the well-grounded perceptual basis without which academic studies may be of very limited value.

Returning to the other purpose of fieldwork, that of making geographical discovery, it would be wrong for any student to infer that, because there are few regions which have not been the subject of exploration, there is limited scope for such discovery. To begin with, geographical phenomena, particularly in the context of this book, i.e. those concerned with agriculture, are constantly undergoing change.

Furthermore, new concepts require the geographer to look again and see familiar things in a new way. New problems challenge the geographer to check again with reality to find solutions which have been overlooked. To keep abreast of changes in the geography of agriculture, and to acquire new ideas and solve new problems as they arise, practical fieldwork leading to original discoveries will be not only always possible, but also very necessary in this as in all other fields.

Fieldwork Visits to Individual Farms

The basic productive unit of a particular system of agriculture, be it farm, ranch, market garden or collective, is the one at which initial fieldwork visits must be undertaken. For convenience, the unit will subsequently be referred to as a farm, and its proprietor, a farmer.

In principle, fieldwork, like all research, should be undertaken without preconceptions which can prevent objectivity in observation. However, the practical problems of knowing what to look for or enquire about, and how best to gain the effective help of the farmer, also require that preliminary research, including study of the available maps and literature of the district, should be undertaken, on the basis of which a relevant, comprehensive and yet brief and uncomplicated questionnaire can be compiled. This questionnaire might well be used for several visits in the same region. Possible preliminary steps in the undertaking of a fieldwork visit may be summarised as follows.

First, it is advantageous for the organisers of such a visit to write to a responsible agricultural agency such as a co-operative, association, farmers' union or possibly the agricultural attaché of an overseas embassy, requesting them, if possible, to aid the students by recommending farmers they deem representative of the more progressive farmers of that region.

Secondly, the organisers should write well in advance, requesting of the farmer a visit, and suggesting a date and time, as well as indicating the number of the proposed party and the purpose of the visit, and including a carefully compiled questionnaire which has been based on the preliminary researches. If this is submitted well in advance it helps the farmer prepare his answers, and gives the students the benefit of his mature thought.

Thirdly, the visit can begin with an interview which may be the occasion of discussion and impromptu questions by the students, based upon and stimulated by the contents of the questionnaire. This interview should be recorded in note-form if not on tape.

The subsequent tour of the holding is now rendered far more purposeful than might otherwise have been the case. More is perceived, and if the farmer is present, misunderstandings can now be clarified, points demonstrated, and previous answers elaborated. This tour should also be the occasion of field observation and note-making, the preparation of field-sketches and supplementary photographs, soil sampling, etc. Such records should be carefully annotated with accurate map references for the locations of soil samples, the points from which observations were made and the directions of such observations. Records should also be kept of the reel and number of each photographic exposure taken in support of the notes and sketches.

Finally, while they are fresh in the memory, the records of the interview, the field-notes, sketches and photographs, in conjunction with earlier research notes and available maps, should all be used to compile an effective farm-study. If these records have been carefully annotated in the field, there should be very little difficulty experienced in cross-checking and building up the complementary sources into a coherent whole. It is often a worthwhile precaution to send a copy of the finished study to the farmer for his perusal and comments.

Questionnaires

A moment's consideration will make it clear that it is practically impossible to design a questionnaire appropriate to all farms within a region, and certainly impossible to design one of universal application. The two included below are therefore intended merely as guides to the student in planning his own. They are reasonably comprehensive in their coverage, yet brief, simple to answer and unambiguous.

QUESTIONNAIRE FOR A "MIXED" FARMER

1. What is the name of the owner, farmer or manager?
2. What is the name and address of the farm?
3. Indicate the position of the farm by its latitude and longitude, distance from and position relative to several nearby towns, or similar.
4. What is the size (in hectares) and extent (distance from north to south, and east to west) of your holding, and is it compact or fragmented?
5. If compact, can you please provide me with a sketch-map, or photocopy of the estate map, showing fields with areas, over-all dimensions, direction of north, scale (approx.), adjacent roads, buildings, etc.?
6. What is the average size of your fields?
7. Is the farm privately owned or tenanted? Does it belong to any co-operative, association, machine syndicate or similar?
8. What are the geology, relief, drainage and soil(s) found in the region around the farm? Are there any contrasts in these from one part of the farm to another? Have you any copies of recent soil analyses?
9. How do the above details affect the agricultural operation of the farm?

10. How long has the farm existed on this site, and what are the farm-house and buildings like?
11. Do you know how the farm-holding and the operation of it have changed over the period since its establishment?
12. For the various kinds of livestock kept please indicate:

Breed of animal	Over-all number of animals	Breakdown of over-all number into "categories"				
		Breeding females	Breeding males	Yearling females	This year's young	Fat beasts
e.g. Ayrshires	106	40 cows	2 bulls	15 heifers	34 calves	15 steers

	Uses	Markets supplied	Seasonal pattern of management
e.g.	Milk	Co-op market at ...	January to June, calving
	Beef	Fatstock market at ...	June to August, markets, etc.

13. For the various kinds of crop grown indicate:

	Area	Species and variety	Yields per ha	Crop grown for
e.g.	40 ha	Wheat–Vestar	5,000 kg/ha	Human consumption

	Markets supplied	Seasonal pattern of management
e.g.	Rank-Hovis, Hull	February, sowing
		April, spraying
		August to September, harvest

14. Indicate the soil dressings appropriate to different kinds of soil/crop.
15. What rotational practices do you employ on the arable area?
16. Have you any other economic resources on the farm such as mining, woodland, fishing or shooting rights, bee-keeping, tourism?
17. Please give the following details of your labour force.
    (a) Total number of people employed, including members of family.
    (b) Number of these in own family (including self). Give details.
    (c) Of the remainder, give numbers of: (i) male and female, (ii) part-time and full-time, (iii) skilled and unskilled, (iv) specialised and general.
    (d) Numbers of casual, seasonal or contractual workers employed, or work done on a reciprocal basis between neighbours.
18. Mechanisation and plant.
    (a) Describe your buildings and installations.
    (b) Describe the more important machines and implements held on the farm.
    (c) Indicate other items you hire or borrow from neighbours as required.
    (d) Do you belong to a machine syndicate, or do you "hire out" any machines to other farmers?
19. Give a detailed breakdown of the sources of your income by percentage, for example, beef 60 per cent, milk 20 per cent, potatoes

5 per cent, tourism 5 per cent, government subsidies 10 per cent. *Note that no money values are required.*

20. What current economic or political changes are affecting farming generally, and your farm in particular? How do you assess the future prospects of your holding?

When completed, please return to ... [Name and address supplied].
Thank you for your help,
Yours sincerely,
[Signature].

QUESTIONNAIRE FOR A MARKET GARDENER

[Most of the questions from the first questionnaire are appropriate down to number 11, although the words "market garden" are best substituted for "farm" in a number of places. Question 12 is probably not appropriate in most regions. Question 13 can be rendered more relevant as follows.]

13. For the various kinds of crop grown indicate:

Area given to crop	Species and variety	Yield per unit area	Use crop put to
e.g. 15 ha	Potatoes— King Edwards	1 tonne per ha	Best for human use, rest for fodder

Markets supplied	Pattern of seasonal management
e.g. Cannery at Local farms	March planting June/September, lifting

[Questions 14 and 15 remain quite relevant, but 16 could also be modified to read as follows.]

16. Have you any other economic resources on the holding such as woodland, bee-keeping, or the sale of produce direct to the public?

[Again, question 17 requires no changes, but 18 and 19 might well be rendered as follows.]

18. Mechanisation and plant.
    (*a*) List your main items of equipment such as rotivators, tractors, seed-drills, lifters or pickers, trailers, etc.
    (*b*) For each of your greenhouses, storage rooms, etc. give specifications such as area in m², construction, heating system employed, automatic devices such as humidisers, etc.

19. Give a detailed breakdown of your sources of income by percentage, for example, potatoes 20 per cent, peas and beans 25 per cent, salad vegetables 25 per cent, soft fruit 30 per cent. *Note that no money values are required.*

[Apart from substituting "market garden" for "farm" as before, the remainder of the questionnaire needs no changes.]

Advantages and shortcomings of such studies

As already indicated, preparation of a farm-study gives the student an insight into the practical character and real problems associated with a farm that is representative of a system, thereby investing the generalised accounts with reality. Whole misconceptions can be corrected in a very short time. Nevertheless, care must be taken in choosing representative farms for study. In a small region, a study of several farms may help to distinguish features in common from those peculiar to one. Farmers may also be asked the ways in which their farms differ from those of their neighbours. There is virtue in asking the specialised organisation which first recommended the farm concerned to select only typical establishments. Careful examination of regional texts will usually help the student to identify the characteristics of agriculture throughout that region. It is then that one is strongly aware of how much the studies in the field have amplified the general account and invested it with fuller meaning.

Presentation, Collation and Analysis of Field Data

There are two main uses to which individual farm-studies may be put; they may be used to throw light on the nature of agriculture generally

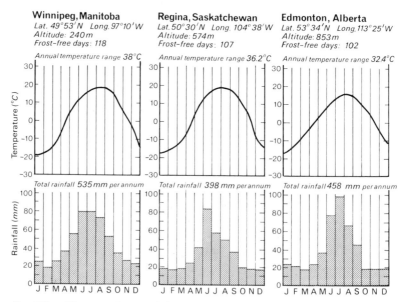

FIG. 108.—*Climatographs, straight line and bar graphs.* To illustrate the climates of three towns in the Canadian Prairie Provinces. (*See also* Fig. 71 in Chapter IX for a bar graph.)

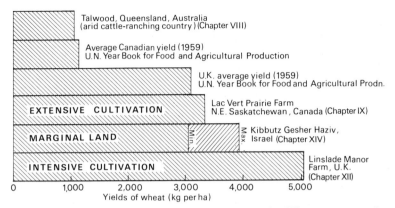

FIG. 109.—*Variations in wheat yields in different areas, under different systems, and at different times.*

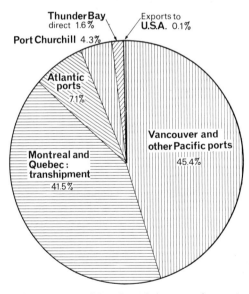

FIG. 110.—*Pie-graph: percentages of Canadian wheat exports from various ports. (See also Fig. 68 in Chapter IX.)*

or a particular agricultural system, or they may be used to illustrate and amplify the general character of a region. In this book it is relevant to consider briefly the ways in which this can be done, but for fuller details of the methods demonstrated here and of many others of equal

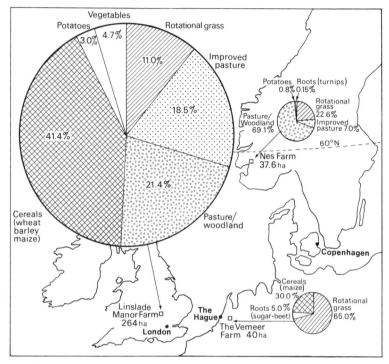

FIG. 111.—*Pie-graphs displaying comparative land-uses of three farms described in Chapter XII.* Over-all size of each graph relates to size of farm.

value the student is recommended to consult books listed in the bibliography, or else those specialising in a consideration of statistical presentation and interpretation, or in simple cartographic techniques for the geographer.

Figures 108–12 have been designed to show just some possible ways in which the data collected for the various chapters of this book could be presented in the form of simple graphs or diagrams or on maps in order to draw attention to causal relationships for farms within a region or within a system, that is, how differences or similarities between them are related to external factors operating upon them.

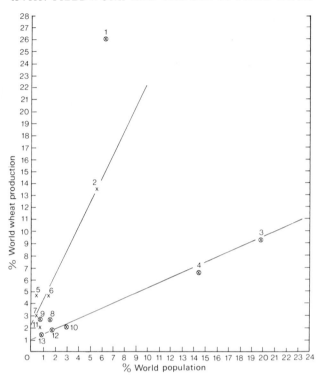

FIG. 112.—*Scatter diagram (based upon Table 5 in Chapter IX).* Shows some correlation between scale of wheat production and share of world population, and whether a country exports or imports. Exporters tend to be large producers and small consumers, importers vice versa. Correlation for importers tends to be weaker (*see* U.S.S.R.) because other factors, e.g. dependability of harvests or per capita consumption levels, may interpose. Crosses indicate surplus/exporters, circled crosses deficit/importers.

Key

1. U.S.S.R.	8. Italy
2. U.S.A.	9. Turkey
3. China	10. Pakistan
4. India	11. Argentina
5. Canada	12. West Germany
6. France	13. Poland
7. Australia	

Spearman's correlation coefficient

One accurate way to establish the degree of correlation which exists between two sets of observations is by the statistical device known as Spearman's correlation coefficient. The formula is as follows:

$$r = \frac{1 - 6(\Sigma d^2)}{n^3 - n}$$

where r = the correlation coefficient
  n = the number of objects tested
  d = the difference in rank order
  Σd^2 = the sum of all the differences squared

TABLE 21

Correlation between scale of wheat production and share of world population

1 Country	2 Percentage of world wheat production	3 Rank order	4 Percentage of world population	5 Rank order	6 d (difference between columns 3 and 5)	7 d²
U.S.S.R.	26.0	1	6.54	3	2	4
U.S.A.	13.3	2	5.53	4	2	4
China	9.0	3	20.00	1	2	4
India	6.3	4	14.20	2	2	4
Canada	4.7	5	0.55	12	7	49
France	4.6	6	1.40	8	2	4
Australia	3.0	7	0.32	13	6	36
Italy	2.8	8	1.50	7	1	1
Turkey	2.8	9	0.90	9	0	0
Pakistan	2.0	10	3.00	5	5	25
Argentina	2.0	11	0.64	11	0	0
West Germany	1.9	12	1.66	6	6	36
Poland	1.5	13	0.90	9	4	16
						$\Sigma d^2 = 183$

Table 21 is based on Table 5 in Chapter IX. Using the figures in Table 21 in Spearman's formula:

$$r = 1 - \frac{(6 \times 183)}{2{,}197 - 13}$$
$$r = 1 - \frac{1{,}098}{2{,}184}$$
$$r = 1 - 0.503$$
$$r = 0.497$$

The value of r always lies between $+1$ and -1. The closer r is to $+1$ or -1, the higher the correlation. In the above example, where $r = +0.497$, there is seen to be a moderate positive correlation between the scale of wheat production and the share of world population. This result is confirmed by the scatter diagram in Figure 112.

How to view Farm-studies in their Wider Regional Context

It remains here to consider ways in which farm-studies can be related to fieldwork undertaken in a larger regional context.

Although it was not relevant to the systematic view of agriculture adopted in this book, it would nevertheless be advantageous to a student seeking to relate farm-studies to a larger regional setting to investigate such ancillary agricultural industries and services as university research institutes, artificial insemination stations, various kinds of food-processing plants, manufacturers of fertiliser or machinery, veterinary practices, etc.

Once secured, the isolated studies can most effectively be integrated into a regional framework by means of careful examination of all available maps, and the undertaking of a personal agricultural land-use survey. As suggested (*see* p. 302), map study should take place both preliminary to and as part of the fieldwork. Maps used should include the largest available, scale 1 : 10,000 if possible (for the U.K., the Ordnance Survey; elsewhere, its nearest equivalent). More specialised information can be sought on a geological sheet and a land utilisation sheet (in the U.K., the second edition Land-use Survey edited by Alice Coleman is preferable), while a map of agricultural types to be found in the region under consideration will also be helpful. Students undertaking work elsewhere than in the U.K. should enquire about maps published by a government or geographical agency there, since a large-scale official publication is very much preferable to a small-scale commercial publication which may be incomplete, inaccurate or out of date.

All these sources should be consulted and compared for the light such comparison is able to throw upon variations in agricultural practice and land-use in quite a small area, even within a single farm-holding. As shown in studies such as those of Felen Rhyd Fach Farm in Chapter VIII, and of Nes and Linslade Manor Farms in Chapter XII, variations in land-use may be based upon such factors as geological contrasts, differences in altitude, liability to flooding or otherwise, or differences in distance between land and farmhouse or communications.

As part of the personal agricultural land-use survey, the student should take soil samples which can be tested mechanically and chemically, and which should be noted at the appropriate locations on the large-scale blank map outline which should be traced or photocopied from the appropriate large-scale map sheet (e.g. a 1 : 10,000 Ordnance Survey map in the U.K.).

The first step in undertaking such a survey, however, is to obtain permission from all the landowners involved. If this is not done, or if permission is not given, walking should be confined entirely to roads or accredited rights of way. Care must be taken not to damage crops, frighten or endanger livestock, leave gates open, or cause any inconvenience to the farming community whose working place this is.

The aim is to produce a detailed field-to-field record of land-use and other relevant observations, e.g. geological outcrops, which will be contemporary with, and therefore complementary to, the farm-studies. On the basis of pencilled instructions, the sheet can later be coloured to conform roughly to the official land utilisation survey sheet, if one is available. In the U.K., the first edition by L. D. Stamp, and particularly the second edition by Alice Coleman, contain much detail irrelevant to the general purposes of the average student of agricultural geography. (Both classifications can be consulted in Leslie Symons, *Agricultural Geography*.)

It is suggested that the student simply identifies those buildings directly relevant to agriculture by annotation, e.g. "Sugar-beet factory", "A. I. Stn." or "Co-op. Egg Pkg. Stn." A simple colour code might be:

1. *light brown* for arable, including all kinds of crops, and also rotational fallow;
2. *Light green* for grass, including grass with scrub or rushes;
3. *dark green* for any type of woodland;
4. *purple* for horticulture—vegetables, flowers or soft fruit in the open or under glass;
5. *purple stripes* for commercial orchard or hard-fruit production.

Notes can be added on the basis of information offered by farmers or from observation in passage over the area concerned, e.g. "15 Friesian cows with calves", "wheat being harvested" or "conifers, Forestry Commission".

Finally, the dates during which the survey was undertaken can be appended to indicate that the survey was correct at that time.

Comparison of this record with an official land-use sheet may reveal some interesting comparisons and contrasts, for example:

1. permanent land-use changes which have occurred since the official survey, e.g. erection of greenhouses, clearance of woodland, or new roads;
2. changes reflecting seasonal variations in land-use, e.g. standing crops replaced at harvest-time by grazing of livestock;
3. changes in the species, breeds and numbers of livestock kept, although these are not necessarily significant unless observed to be general over a large area, confirmed in conversation with farmers, or satisfactorily explained by the student's own observations, e.g. a switch from dairying to beef production occasioned by E.E.C. pressure, increased wintering of yearling ewes on upland farms, or land neglect on the periphery of a large city occasioned by imminent land developments.

Having related the farms to the over-all agricultural character of the region they are situated in, it remains of course to consider the wider economic and cultural aspects of the same region. This, however, is outside the scope of the present book.

Bibliography

Books

Abler, R., Adams, J. S. and Gould, P., *Spatial Organisation*, Prentice-Hall, 1971.

Alexander, J. W., *Economic Geography*, Prentice-Hall, 1963.

Bhalerao, M. M., *Agricultural Co-operation in India*, Plunkett Foundation for Co-operative Studies, Occasional Paper No. 35, 1970.

Bolton, T. and Newbury, P. A., *Geography Through Fieldwork: Book 2*, Blandford Press, 1968.

Bonner, Arnold, *British Co-operation*, Co-operative Union, Manchester, revised edition edited by Brian Rose, 1970.

Bowen-Jones, H., Dewdney, J. C. and Fisher, W. B., *Malta—Background for Development*, University of Durham, 1961.

Bradford, M. G. and Kent, W. A., *Human Geography: Theories and their Applications*, Oxford University Press, 1977.

Chisholm, M., *Rural Settlement and Land Use*, Hutchinson, 2nd edition, 1968. *Geography and Economics*, Bell's Advanced Economic Geographies, 1966.

Christaller, W., *Central Places in South Germany*, translated by C. W. Bashen, Prentice-Hall, 1966.

Cole, J. P., *Geography of World Affairs*, Penguin, 5th edition, 1979.

Criden, Y. and Gelb, S., *The Kibbutz Experience—Dialogue in Kfar Blum*, Schocken Books, New York, 1974.

Digby, M., *Select Readings in Agricultural Co-operation*, Plunkett Foundation for Co-operative Studies, 1967.

Forde, C. Daryll, *Habitat, Economy and Society*, Methuen, 1934, 1968.

Found, W. C., *A Theoretical Approach to Land-Use Patterns*, Edward Arnold, 1971.

Haggett, P., *Geography—a Modern Synthesis*, Harper & Row, 1975.

Hall, P. (Ed.), *Von Thünen's Isolated State*, Pergamon Press, 1966.

Hoover, E. M., *The Location of Economic Activity*, McGraw-Hill, 1948.

Isard, W., *Location and Space Economy*, Wiley, 1956.

Jones, E., *Human Geography*, Chatto and Windus, 1969.

Knapp, Joseph G., *An Analysis of Agricultural Co-operation in England*, Agricultural Central Co-operative Association Ltd., London.

Knudsen, P. H., *Agriculture in Denmark*, Agricultural Council of Denmark, Copenhagen, 1977.

Lösch, A., *The Economics of Location*, Yale University Press, 1954.

Lowry, J. H., *World Population and Food Supply*, Edward Arnold, 1977.

McCready, K. J., *The Land Settlement Association—its History and Present Form*, Plunkett Foundation for Co-operative Studies, Occasional Paper No. 37, 1974.

Morley, J. A. E., *British Agricultural Co-operatives*, Hutchinson, 1975.

Nobbs, J., *Advanced Level Economics*, McGraw-Hill, 2nd edition, 1976.

Nye, P. H. and Greenland, D. J., *The Soil under Shifting Cultivation*, Commonwealth Agricultural Bureaux, Farnham Royal, 1960.

Parker, W. H., *Anglo-American*, a Systematic Regional Geography, University of London Press, 1965.

Post, Laurens Van Der, *Lost World of the Kalahari*, Hogarth, 1958, 1961.

Robinson, H., *Geography for Business Studies*, Macdonald & Evans, 3rd edition, 1979.

Samuelson, Paul, *Economics—Introductory Analysis*, McGraw-Hill, New York, 3rd edition, 1955.

Sargent, M. J. and Doherty, N. M., *Agricultural Co-operation in Practice*, University of Bath, 1977.

Stanlake, C. F., *Introductory Economics*, Longman, 3rd edition, 1976.

Symons, Leslie, *Agricultural Geography*, Bell's Advanced Economic Geographies, 1967.

Tarrant, J. R., *Agricultural Geography*, David & Charles, 1974.

Tidswell, W. V., *Pattern and Process in Human Geography*, University Tutorial Press, 1976.

Unstead, J. F., *World Survey from the Human Aspect*, a Systematic Regional Geography, University Tutorial Press, 1965.

Webster, F. H., *Agricultural Co-operation in Denmark*, Plunkett Foundation for Co-operative Studies, Occasional Paper No. 39, 1973.

Wooldridge, S. W. and East, W. G., *The Spirit and Purpose of Geography*, Hutchinson University Library, 3rd edition, 1966.

Worsley, Peter, *Two Blades of Grass: Rural Co-operatives in Agricultural Modernisation*, Manchester University Press, 1971.

Journals and Other Publications

Beeley, B. W., Report on "The Farmer and Rural Society in Malta", Durham, 1959.

Grotewald, A., "Von Thünen in retrospect", *Economic Geography*, Vol. 35, pp. 346–55.

Jonasson, O. "The agricultural regions of Europe", *Economic Geography*, Vol. 1 (1925), pp. 277–314.

Sauer, C. O., "Agricultural origins and dispersal", *American Geographical Magazine*, New York, 1952.

"The State farm at Bala", article describing the reindeer ranch of the Yakutsk peoples of north-east Siberia, *Geographical Magazine*, March 1968, pp. 961–7.

Yearbook of Agricultural Co-operation, 1976, Plunkett Foundation for Co-operative Studies.

Yearbook of Agricultural Co-operation, 1977, Plunkett Foundation for Co-operative Studies.

Other Sources of Information

Below are listed some useful addresses for obtaining further information on the main countries discussed in the text.

Australia

Beef Cattle Husbandry Branch, Department of Primary Industries, P.O. Box 300, Goondiwindi, Queensland 4390

Registrar of Primary Producers' Industries (Agricultural Section), William Street, Brisbane, Queensland 4000

The Co-operative Federation of Australia, Box 5211AA, G.P.O. Melbourne, Victoria, 3001

Registrar of Co-operative and Other Societies, Box 8956, G.P.O. Brisbane, Queensland

The Co-operative Federation of Queensland, G.P.O. Box 2066, Co-op House, 436 Queen Street, S. Brisbane

Canada

Agriculture Canada, Ottowa K1A 0O5

Co-operative Union of Canada, 237 Metcalfe Street, Ottowa K2P 1RZ

Canadian Co-operative Wheat Producers, Albert Street, 8 Victoria Avenue, Regina, Saskatchewan

The Palliser Wheat Growers' Association, No. 219, 3806 Albert Street, Regina, Saskatchewan S4S 3RS

Saskatchewan Wheat Pool, Head Office, Regina, Saskatchewan

Canada Grains Council, Suite 500, 177 Lombard Avenue, Winnipeg, Manitoba R3B 0W5

Denmark

The Agricultural Council Information Section, Axelborg, Axeltorv 3, 1609, Copenhagen V

Federation of Danish Co-operative Societies, Vester Farimagsgade 3, 1606, Copenhagen V

Federation of Danish Farmers' Unions, Axelborg 4, Sal, Vesterbrogade 4A, 1620, Copenhagen V

The Danish Smallholders' Unions, Vester Farimagsgade 6, 1606, Copenhagen V

The Danish Agricultural Marketing Board, Axelborg, Axeltorv 3, 1609, Copenhagen V

E.E.C.

Commission of the European Communities (London Office), 20 Kensington Palace Gardens, London W8 4QQ

C.O.P.A. (Committee of Professional Agricultural Organisations), Rue de la Science, 23/25 Bte, 3, 1040 Brussels

C.O.G.E.C.A. (General Committee of Agricultural Co-operation), Rue de la Science, 23/25 Bte, 3, 1040 Brussels

International Federation of Agricultural Producers, 1 Rue d'Houteville, Paris X

Israel

Ministry of Labour (Co-operative Department), Herze Street, Jerusalem

Farmers' Federation of Israel, Beit Hackarim, P.O. Box 209, Tel Aviv

I.N.V.A. (Central Co-operative for the Marketing of Agricultural Produce), P.O. Box 265, Tel Aviv

Malaysia

Ministry of Agriculture and Co-operatives, Jalan Swettenham, Kuala Lumpur

Co-operative Union of Malaysia, Peti Surat 685, Kuala Lumpur

Farmers' Organisation Authority, Block B, Tingkat Dua, Dompleks Pejabet Damansara, Jalan Dungun, Dawansara Heights, Kuala Lumpur 23–04

Federal Agricultural Marketing Authority, Wisma Yan, Blocks E and F, Jalan Salangor, Petaling Jaya, Selangor

Netherlands

Ministry of Agriculture and Fisheries, 4, Le V. D. Boschstraat, The Hague

National Co-operative Council, Groenhovenstraat 3, The Hague

C.E.B.E.C.O. [national co-operative purchasing and selling society for farming and horticulture], Post Bus 182, Rotterdam

C.O.V.E.C.O. [co-operative association of the central organisation of cattle marketing and processing co-operatives], Bergstraat 35, Arnhem

Norway

Ministry of Agriculture, P.O. Box 8007 Dep, Oslo 1

Central Federation of Agricultural Co-operatives, P.O. Box 3746, Gamlebyen, Oslo

Norwegian Farmers' Union, Schweigaandsgate 34, Oslo

Norwegian Smallholders' Union, Schweigaandsgate 34, Oslo

Norwegian Sales Centre for Dairy Products, Breigate 10, Oslo

U.K.

Ministry of Agriculture, Fisheries and Food, Whitehall Place, London SW1

Agricultural organisation societies, including:

Agricultural Co-operation and Marketing Services, Agriculture House, 25–31 Knightsbridge, London SW1X 7NJ

Scottish Agricultural Organisation Society Ltd., Claremont House, 18–19 Claremont Crescent, Edinburgh EH7 4JW

Welsh Agricultural Organisation Society Ltd., P.O. Box 7, Brynawal, Great Darkgate Street, Aberystwyth SY23 1DR

Ulster Agricultural Organisation Society Ltd., 20 High Street, Portadown, Craigavon BT62 1HU

National Farmers' Union, Agriculture House, 25–31 Knightsbridge, London SW1X 7NJ

Milk Marketing Board, Thames Ditton, Surrey

Potato Marketing Board, 50 Hans Crescent, London SW1X 0NB

British Wool Marketing Board, Vakmills, Station Road, Clayton, Bradford, West Yorkshire BD14 6JD

U.S.A.

The U.S. Department of Agriculture, Foreign Agricultural Service, Office of Agricultural Attaché, American Embassy, Grosvenor Square, London W1A 1AE

The National Federation of Grain Co-operatives, 711 14th Street N.W., Washington D.C. 20005

Director of Information, Kentucky State Department of Agriculture, Capitol Annex Building, Frankfort, Kentucky 40601

California Fruit Exchange, P.O. Box 15498, Sacramento, California 95814

General Information on Co-operative Agriculture

Plunkett Foundation for Co-operative Studies, 31 St. Giles, Oxford OM1 3LF, England.

Index